Praise for
Demand and Supply Integration

"Based on his nearly two decades of supply chain research and consulting, Dr. Moon succinctly describes the what, why, and how of Demand and Supply Integration in this book. It is a powerful reference guide for every supply chain leader and practitioner."

—**Reuben E. Slone**, Senior Vice President, Supply Chain, Walgreens

"Mark Moon finds the right balance between academic rigor and practitioner relevance as he moves the focus of demand and sales forecasting from the historical realm of statistics to the realm of strategic decision making. This book continues building on the proven best practices methodology previously inaugurated by his mentor and colleague, Tom Mentzer. This is a well-done, forward-thinking, practical guide for business professionals who want to improve their forecasting processes."

—**Dwight Thomas**, Retired, Former Supply Chain Planning Manager for Alcatel-Lucent Technologies

"Mark Moon has provided excellent insight in defining the importance of integrating demand and supply into a true business process and demonstrating the top and bottom line impact. This is a fresh viewpoint that can help organizations add value to their current process or to provide guidance for a new process."

—**Edward Guzowski**, Director, SIOP, Eaton Corporation

"Moon is one of the leading minds on the rapidly evolving subject of Demand and Supply Integration. In this book he grounds thought-provoking insights on the subject with the abundant use of current, real-world examples from a variety of different industries. A must-read for both the practitioner and the academic."

—**Nikhil Sagar**, Vice President, Inventory Management, Fortune 100 Retailer

"Brilliantly organized and an incisive examination of demand-supply integration (DSI) within an organization and across the supply chain. DSI is what S&OP is supposed to be. Professor Moon carefully portrays an *ideal* DSI process, the stages a company passes through to get there, and the aberrant actions and policies that can undermine it. A must-read for planners and managers and a valuable complementary text in MBA business-forecasting courses."

—**Len Tashman**, Editor, *Foresight: The International Journal of Applied Forecasting*

"Sales or demand forecasting is an enigma for most. Mark Moon, in a very clear and straightforward way, elucidates the discipline and its important business roles. These include not only driving an efficient supply chain, but also its value as a strategic enterprise asset when properly utilized to inform all planning processes, both short and long term."

—Dave Pocklington, Vice President, Sales Forecasting & Analytics, Global Strategic Planning, Amway Corporation

Demand and Supply Integration

Demand and Supply Integration:

The Key to World-Class Demand Forecasting

Mark A. Moon

FT Press

Vice President, Publisher: Tim Moore
Associate Publisher and Director of Marketing: Amy Neidlinger
Executive Editor: Jeanne Glasser Levine
Editorial Assistant: Pamela Boland
Operations Specialist: Jodi Kemper
Marketing Manager: Megan Graue
Cover Designer: Chuti Prasertsith
Managing Editor: Kristy Hart
Senior Project Editor: Lori Lyons
Copy Editor: Paula Lowell
Proofreader: Kathy Ruiz
Indexer: Erika Millen
Senior Compositor: Gloria Schurick
Manufacturing Buyer: Dan Uhrig

This book is sold with the understanding that neither the author nor the publisher is engaged in rendering legal, accounting, or other professional services or advice by publishing this book. Each individual situation is unique. Thus, if legal or financial advice or other expert assistance is required in a specific situation, the services of a competent professional should be sought to ensure that the situation has been evaluated carefully and appropriately. The author and the publisher disclaim any liability, loss, or risk resulting directly or indirectly, from the use or application of any of the contents of this book.

FT Press offers excellent discounts on this book when ordered in quantity for bulk purchases or special sales. For more information, please contact U.S. Corporate and Government Sales, 1-800-382-3419, corpsales@pearsontechgroup.com. For sales outside the U.S., please contact International Sales at international@pearsoned.com.

Printed in the United States of America
First Printing January 2013

ISBN-10: 0-13-308801-4
ISBN-13: 978-0-13-308801-4

Pearson Education LTD.
Pearson Education Australia PTY, Limited.
Pearson Education Singapore, Pte. Ltd.
Pearson Education Asia, Ltd.
Pearson Education Canada, Ltd.
Pearson Educación de Mexico, S.A. de C.V.
Pearson Education—Japan
Pearson Education Malaysia, Pte. Ltd.

Library of Congress Cataloging-in-Publication Data

Moon, Mark A., 1955-
    Demand and supply integration : the key to world-class demand forecasting / Mark A. Moon.
      p. cm.
    ISBN-13: 978-0-13-308801-4 (hardcover : alk. paper)
    ISBN-10: 0-13-308801-4 (hardcover : alk. paper)
    1. Supply and demand—Forecasting. 2. Management. I. Title.
    HB801.M6923 2013
    338.5'210112--dc23
                                2012043608

To Tom Mentzer, who got me started in this stuff.

To my sons, Colin and David.
I couldn't be prouder of you both.

To Carol, who has made me happier
than I could ever deserve.

Contents at a Glance

Contents

Acknowledgments

I would like to acknowledge three categories of people and organizations as being indispensable contributors to this book: the colleagues with whom I've worked, the companies from which I've learned about forecasting and DSI, and others who have played a very special role in the writing of this book.

First are the colleagues with whom I've worked. I've already mentioned Tom Mentzer, and he is the first person whom I must acknowledge. Without him, none of this happens. He taught me about forecasting and he opened the doors that needed to be opened to get access to the companies. His tragic death in 2010 is still felt by all of us at the University of Tennessee. His were very big shoes, both literally and figuratively. He was one of the worst golfers I ever had the joy of playing with, and the most generous-spirited man I ever met. I miss you every day, Tom.

After Tom, there were other faculty colleagues who participated in audits, consulting projects, and executive teaching. Ted Stank, Funda Sahin, and Ken Kahn were wonderful colleagues with whom I have worked. Of particular note is my dear friend, Paul Dittmann. We've traveled the world together, and everything I've learned about inventory management, and most of what I've learned about supply chain management, I've learned from this insightful, hard-working, and dedicated professional. In addition to the faculty colleagues were the doctoral students with whom we worked. We never could have collected all that data from the companies in our audit database without their excellent participation and contribution. Those who come to mind include Carlo Smith, John Kent, Nancy Nix, Brian Fugate, Beth Davis-Sramek, Barbara Marshall, Andy Artis, Marcel Zondag, Melinda Jones, Michelle Bobbitt, Michael Garver, Cliff DeFee, and Shay Scott. All have gone on to various faculty positions around the world, and they were all fun to work with and drink some beer with at the end of those long days of interviews. Particular acknowledgement must go to "Mark's Angels"—Donna Davis, Teresa McCarthy-Byrne, and Susan Golicic. Altogether, those three remarkable women worked with me on about 10 different audits, in various configurations. They abandoned me in Fort Worth, partied with me on South Beach, picked out lingerie in Winston-Salem, and even tried to find me (unsuccessfully) in Dublin. But that's another story. The four of us became dear friends, and we did a lot of great work together learning about forecasting and demand/supply integration.

Also deserving of acknowledgement are the companies with which we worked. In a later chapter of this book, I name all 42 companies that have participated in the audit research, which led to so many of the insights contained in this book. In addition to these 42, dozens of companies that have been members of the Sales Forecasting Management Forum and the Supply Chain Management and Strategy Forum at the University of Tennessee have provided insight and guidance over the years. A few of these professional colleagues stand out and deserve special mention:

- John Hewson, from Eastman Chemical, was the very first forecasting champion we met. His tragic and sudden passing in the early 2000s left a gaping hole in that company.
- Ken Carlson from Deere and Company. Ken taught us about creating a culture of engagement, and about taking the forecasting process global. Ken was the wizard of forecasting.
- Dwight Thomas from Lucent Technologies. Dwight led a global team of forecasters for many years, and taught us about how to engage a worldwide sales organization in a highly complex forecasting task.
- Dave Pocklington from Amway. Dave and his team helped us to understand that forecasting can be more than a tactical function, but rather, can be a corporate function that contributes to the strategic decision-making of the firm. In 16 years, I've still not seen a company that does it better than Amway. Dave and

his team from Amway felt so strongly about what we've done in forecasting that they endowed a scholarship at the University of Tennessee that is awarded each year to an undergraduate student who is interested in forecasting and demand planning.

- John Hellriegel from Honeywell. John has helped me to organize and articulate my ideas about forecasting excellence. We've traveled the world together teaching forecasting to hundreds of individuals at Honeywell. We've even eaten the hairy crab in Shanghai.

Many other professional colleagues have been a part of my learning journey, and I thank them all for their kindness and generosity.

Finally, I acknowledge the others who have contributed in special ways to the writing of this book. Chad Autry, my colleague at UT, helped make contact with my publisher. Jeanne Glasser Levine, from Financial Time Press, has been a cheerleader and timekeeper and helped immeasurably in getting this book completed. My sons, Colin and David, and my step-daughter, Lauren, have cheered me on. And most importantly, my beautiful wife, Carol, has made it possible to spend many hours in front of the computer and has given me the love and support to do something that I never thought I could do—start with a blank piece of paper, and write a book. You're the bomb.

About the Author

Dr. Mark A. Moon is an Associate Professor of Marketing and Head of the Department of Marketing and Supply Chain Management at the University of Tennessee, Knoxville. Prior to joining the UT faculty in 1993, Dr. Moon earned his Ph.D. from the University of North Carolina at Chapel Hill. He also holds MBA and BA degrees from the University of Michigan in Ann Arbor. Dr. Moon's professional experience includes positions in sales and marketing with IBM and Xerox. He teaches at the undergraduate, MBA, and Executive MBA, and Ph.D. levels, and teaches demand planning, forecasting, and marketing strategy in numerous executive programs offered at the University of Tennessee's Center for Executive Education. Dr. Moon's primary research interests are in demand management, sales forecasting, buyer-seller relationships, and demand/supply integration (or sales and operations planning). He has published in the *Journal of the Academy of Marketing Science, International Journal of Forecasting, Supply Chain Management Review, Foresight, Journal of Personal Selling and Sales Management, Journal of Business Forecasting, Journal of Marketing Education, Marketing Education Review, Business Horizons, Industrial Marketing Management, Journal of Marketing Theory and Practice,* and several national conference proceedings. Dr. Moon is also the author, along with Dr. John T. (Tom) Mentzer of *Sales Forecasting Management: A Demand Management Approach.*

Dr. Moon has consulted with numerous companies on sales forecasting re-engineering projects, including AET Films, AlliedSignal, Amway, Avery Dennison, Bacardi USA, Conagra, Continental Tire, Cooper Tire, Corning, Deere and Company, DuPont, Eastman Chemical, Ethicon, Exxon, Hershey Foods, Lucent Technologies, Maxtor, Michelin, Motorola PCS, OfficeMax, Orbit Irrigation Products, Peerless Pumps, Pharmavite, Philips Consumer Electronics, Sara Lee Intimate Apparel, Smith & Nephew, Union Pacific Railroad, Whirlpool, and Williamson-Dickie. He has also consulted with numerous companies on the topic of supply chain strategy, including Lockheed-Martin, Nissan North America, Johnson & Johnson, Radio Systems Corporation, Cummins Filtration, Tyco, and Winn-Dixie. In addition, Dr. Moon has delivered custom executive education programs, covering topics that include marketing strategy, sales forecasting, demand planning, and sales and operation planning, with numerous companies, including Honeywell, Coca-Cola, Corning, BASF, 3M, Union Pacific Railroad, EdAmerica, Nestle, Orbit Irrigation Products, Sony, American Standard, and CHEP.

Mark was born and raised in Ann Arbor, Michigan. He has two sons: Colin and David. Away from the office, Mark enjoys traveling with his wife, Carol, and golfing with his sons, who now beat him regularly.

Preface

Back in May 1996, I was a (relatively) young assistant professor who had joined the faculty at the University of Tennessee almost three years earlier, following my graduate school years in Chapel Hill, North Carolina. My field was, and is, marketing. I had worked for several years in sales with IBM before going back to school to get my Ph.D., and I was doing research and teaching in various areas of sales force management. That day in May 1996, we had just finished the spring semester, and I was looking forward to a summer of writing and revising articles, playing a little golf, and hanging out with my young sons. I remember so clearly sitting in my office, minding my own business, when the most prominent of our senior faculty stopped by, stuck his head in the office, and asked me a question that changed the course of my career. That senior faculty member was Tom Mentzer, and the question he asked was, "Would you like to do some work with me on forecasting? I've got a company that needs a forecasting audit, and I'm wondering if you'd like to help out." Well, when you're an assistant professor, three years into your career, and someone with the stature of Tom Mentzer stops in and offers you an opportunity to work together, no one in his right mind would say no. So I said, "Sure, Tom. What's forecasting?" He smiled, walked back to his office, and returned a few minutes later with about 20 of the articles he had written on the subject, 20 or so articles that others had written on the subject, and a couple books. He suggested that I read as much of that pile as I could in the next week, because we were due in Kingsport, Tennessee, a week later to begin doing a forecast audit for Eastman Chemical Company.

So Tom and I, along with two of our doctoral students—John Kent and Carlo Smith—spent the next few weeks driving the 90 miles back and forth from Knoxville to Kingsport, and thus began my education into forecasting. We completed our assessment for Eastman Chemical, and then in quick succession completed forecasting audits for DuPont Agricultural Products, Hershey Foods, Michelin Tire, and AlliedSignal Automotive Products. In the 16 years since that day in May, I've learned a lot. Tom and I worked with a lot more companies—34 companies in total—before his tragic and untimely death in 2010. I then joined forces with my colleague Paul Dittmann and others, and audited another eight companies, bringing the total to 42. Along with a variety of co-authors, I've written a lot of articles about forecasting, and joined forces with Tom Mentzer on a second edition of his well-known and well-respected forecasting textbook. I've also taught forecasting and demand planning to a lot of students—undergraduate and MBA students at the University of Tennessee, the Bordeaux Ecole de Management in Bordeaux, France; and working forecasters and demand planners at companies such as Union Pacific Railroad, Corning, Orbit Irrigation Products, BASF, 3M, and Honeywell. I've traveled the world learning about forecasting practices and teaching executive audiences—from various places in the United States, to Canada, Mexico, China, Switzerland, France, the Netherlands, England, Ireland, Singapore, Taiwan, and Belgium. I learned a lot about best practices, and I learned a lot about worst practices.

Interestingly, in the course of that journey, I came to see forecasting in a different light from the light that Tom saw it in. I came to see that demand forecasting in and of itself was not particularly helpful for a company. I came across many companies that were pretty good at forecasting, but they still struggled with their inventories, fill rates, and costs. The reason they struggled with these problems was not that they were not forecasting well. Rather, it was because they were not doing a good job of translating their forecasts into good business decisions. Sales and marketing were not communicating well with their supply chain colleagues, and vice versa. In other words, I came to see that forecasting wasn't the only thing that companies needed to work on. They also needed to work on those integrating processes that facilitate communication between the demand side of the firm (sales and marketing in a manufacturing context, and merchandising in a retailing context) and the supply side of the firm (the supply chain organization, or logistics, procurement, and operations).

This realization fit in extremely well in my home department at the University of Tennessee. I am a proud member of the Department of Marketing and Supply Chain Management. It might seem strange to some people that an academic department in a College of Business Administration would be a combination of marketing and supply chain management. At Tennessee, these two units were put together back in the 1970s for purely cost-saving reasons—save a couple of administrative assistants if you smash two departments together. For many years, those two units didn't have much to do with each other. But once again, I invoke the name of Tom Mentzer, who was an extremely prominent scholar in both marketing and supply chain management. He was at various times in his career the president of *both* the Council for Logistics Management (now the Council for Supply Chain Management Professionals) *and* the Academy of Marketing Science. His larger-than-life personality, and the force of his convictions, helped our department to see the synergy between marketing and supply chain management, and thanks to his leadership, we developed a vision of business practice that we refer to as *Demand and Supply Integration (DSI)*. Tom helped us get started developing this vision, and since his passing, those of us who remain at Tennessee have continued to develop and refine this DSI vision. We've written articles, both academic and practitioner-oriented, that articulate our thoughts about how demand (sales and marketing) and supply (supply chain) need to be integrated through culture, processes, and tools, for the betterment of the enterprise as a whole. This book is meant to articulate an important element of that integration, namely the processes that I refer to as DSI (Demand/Supply Integration) and the demand-side contribution to that DSI process, namely the *Demand Forecast*. In this way, I am attempting to take forecasting out of the realm of statisticians and bring it into the realm of strategic

decision makers. Thus, as the title of this book suggests, Demand/
Supply Integration, or DSI, is the key to excellence in demand forecasting. As I say in later chapters, an accurate forecast and 50 cents
will buy you a cup of coffee. Without the DSI process as the foundation, then the demand forecast isn't worth the paper it's written on.

What This Book Is, and What It Is Not

The following table provides a summary of what you can expect
from this book, and what you cannot expect.

What This Book Is	What This Book Is Not
A guide for those who develop and manage the DSI processes in an organization	A book only about forecasting
A guide for business professionals on how to manage their forecasting processes	A comprehensive textbook about all things forecasting
Helpful advice on how companies measure performance in practice, and how they use those performance measurements to drive behavior and make good business decisions	A detailed discussion of all possible ways to measure forecast accuracy
Guidance on how best to combine the insights that come from examination of historical demand patterns (statistical forecasting) with the insights that come from the qualitative judgments of knowledgeable businesspeople	A detailed discussion of statistical forecasting, along with formulas, assumptions, and tests
Based on 16 years of working with hundreds of forecasting professionals, learning what works and what doesn't	Rigorously "proven" academic theories about forecasting or demand/supply integration

A few summary comments:

- If you're looking for a textbook that teaches statistical forecasting, in all its nuances, this ain't it. Many other excellent books
  have been published that will do that much better than I can.

However, if you're looking for practical advice on how to successfully combine the insights from statistical forecasting with the insights from qualitative judgment, you'll find that here. You'll also find some pitfalls of using the wrong statistical tools, and some of the problems associated with use of simple statistical methods that sometimes do more harm than good.

- In the same vein, if you're looking for a book that articulates every possible way to measure forecast accuracy, this also ain't it. Once again, I focus on practical advice for working forecasting professionals, giving instruction about *how* to measure performance, but also insight about *why* you measure performance.

- This book is written for practicing managers. It is intended to be, above all else, practical and useful for those practicing managers. It's based on what I've seen that works well, and what I've seen not work well. There are no hypotheses tested in this book—just practical advice for managers.

- This book is mostly about forecasting, but *in the context of Demand/Supply Integration*. The first chapter is dedicated to Demand/Supply Integration, and the last chapter brings the reader back to that super-process, but everything in-between is intended to give managers of forecasting processes practical advice on how to do demand forecasting better.

How This Book Is Organized

I begin and end this book with a discussion of demand/supply integration. Chapter 1, "Demand/Supply Integration," articulates the goals of DSI and explains how DSI is different from Sales and Operations Planning, or S&OP. I articulate common aberrations from the ideal state of DSI, and discuss an ideal sequence of subprocesses that need to be in place to support the "super-process" of DSI. Chapters 2 through 7 focus on demand forecasting. In Chapter 2, "Demand

Forecasting as a Management Process," I make the point that demand forecasting is a management process, which must be planned, executed, and controlled like any other management process. In this chapter, I introduce concepts such as the definition of a forecast, the definition of demand, the forecasting hierarchy, and the information technology systems that support forecasting excellence.

Chapters 3 and 4 are dedicated to forecasting techniques. Chapter 3, "Quantitative Forecasting Techniques," is a discussion of quantitative, or statistical, forecasting, with emphasis placed on the "why," rather than on the detailed "how." Chapter 4, "Qualitative Forecasting Techniques," discusses the process of enriching the statistical forecast with input that comes from those individuals who might have insight about how the future might look different from the past. Chapter 5, "Incorporating Market Intelligence into the Forecast," expands the discussion of qualitative insight begun in Chapter 4 by discussing the role that market intelligence plays in the forecasting process, as well as the useful contribution that can come from customer-generated forecasts.

In Chapter 6, "Performance Measurement," I discuss the importance of, and the steps required, to measure forecasting performance. Like any other management process, if you can't measure it, you can't manage it. Chapter 7, "World-Class Demand Forecasting," then describes, in considerable detail, the vision we have developed at the University of Tennessee of world-class forecasting. This chapter allows you to do a self-assessment of your own forecasting processes, and see where you should focus your re-engineering efforts. The book concludes with Chapter 8, "Bringing It Back to Demand/Supply Integration: Managing the Demand Review," where I bring the discussion back the demand/supply integration, or DSI. It covers the demand review, which is the culmination of the forecasting process, and how to prepare for, and effectively manage, that critical piece of DSI.

One brief note on terminology: Throughout this book, I wander back and forth between the pronouns *I* and *we*. Although I am officially the sole author of this book, it's a very collaborative effort. Hundreds of colleagues—ranging from other faculty at the University of Tennessee, to doctoral students, to people at the companies with whom we've worked, to other students—have helped me form the ideas documented in this book. There are times when I feel compelled to use *we*, because so many of these ideas are the results of so much collaboration.

So with all that said, let's get into it. I hope you enjoy this guide through demand forecasting in a demand/supply integration context. If you find things you don't like, or don't agree with, or are simply wrong, let me know. And, Go Vols!

Mark Moon
Knoxville, TN

1

Demand/Supply Integration

One of the companies that participated in the *DSI/Forecasting Audit research* was in the apparel industry. This company, a manufacturer and marketer of branded casual clothing, had very large retail customers that contributed a large percentage of overall revenue. Understandably, keeping these large retail customers in stock was very important to the success of this company. If these retailers' orders could not be filled, then out-of-stock conditions would result, with not only lost sales as the consequence, but also potential financial penalties for failure to satisfy these retailers' stringent fill-rate expectations.

As is the case for many companies in this industry, considerable manufacturing capacity had been offshored to sewing operations in Asia. This strategy helped to keep unit costs down, but it also had a negative impact on the company's responsiveness and flexibility. At the time of our audit, the research team heard about a communication disconnect between the supply chain and the sales organizations at this company. A variety of problems had left the company with significant capacity shortages. Although these problems were solvable in the long run, in the short term, the company was having significant fill-rate problems with some of its largest, most important retail customers. Some of the most popular sizes and styles of clothing were in short supply, and customers were not happy. Supply chain personnel were working hard to address these problems, but in the short term, there was little to be done. Although these supply chain problems were impacting the company's largest, most important customers, personnel from the field sales organization were being incentivized to open new channels of distribution and locate new customers to carry their brands. As one supply chain executive told this story, she said in exasperation, "We're out of stock at Wal-Mart, and they're signing up new customers! What the hell is going on here?"

This example is a classic illustration of what can happen when *Demand/Supply Integration*, or DSI, is not a part of the fabric of an organization. This chapter explores the essence of DSI, distinguishes it from Sales and Operations Planning (S&OP), articulates from a strategic perspective what DSI is designed to accomplish, describes some typical aberrations from the "ideal state" of practice, and describes some characteristics of successful DSI implementations.

The Idea Behind DSI

Demand/Supply Integration (DSI), when implemented effectively, is a *single process* to engage *all functions* in creating *aligned, forward-looking plans* and *make decisions* that will optimize resources and achieve a balanced set organizational goals. Several phrases in the preceding sentence deserve further elaboration. First, DSI is a *single process*. The idea is that DSI is a "super-process" containing a number of "subprocesses" that are highly coordinated to achieve an overall aligned business plan. These subprocesses include demand planning, inventory planning, supply planning, and financial planning. Second, it is a process that engages *all functions*. The primary functions that must be engaged for DSI to work effectively are sales, marketing, supply chain, finance, and senior leadership. Without active, committed engagement from each of the functional areas, the strategic goals behind DSI cannot be achieved. Third, it is designed to be a process that creates *aligned, forward-looking plans and makes decisions*. Unfortunately, when DSI is not implemented well, it often consists of "post-mortems," or discussions of "why we didn't make our numbers last month." The ultimate goal of DSI is business planning—in other words, what steps will an organization take *in the future* to achieve its goals?

Our research has shown that three important elements must be in place for DSI to operate effectively: *culture, process,* and *tools.* An organization's culture must be focused on transparency, collaboration, and commitment to organization-wide goals. Processes must be clearly articulated, documented, and followed to ensure that all planning steps are completed. Effective tools, normally thought of as information technology tools, are also needed to provide the right information at the right time to the right people.

How DSI Is Different from S&OP

Many authors have, over the last 20+ years, written about Sales and Operations Planning (S&OP), and to some, the earlier description of DSI might sound like little more than a rebranding of S&OP. Unfortunately, S&OP has a bit of a "bad name," thanks to ineffective process implementation. In our observation of dozens of S&OP implementations, we've seen several common implementation problems that have contributed to a sense of frustration with the effectiveness of these processes.

First, S&OP processes are often tactical in nature. They often focus on balancing demand with supply in the short run, and turn into exercises in flexing the supply chain, either up or down, to respond to sudden and unexpected changes in demand. The planning horizon often fails to extend beyond the current fiscal quarter. With such a tactical focus, the firm can miss out on the chance to make strategic decisions about both supply capability and demand generation that extend further into the future, which can position the firm to be proactive about pursuing market opportunities.

Second, S&OP process implementation is often initiated, and managed, by a firm's supply chain organization. In our experience, these business-planning processes are put into place because supply chain executives are "blamed" for failure to meet customer demand in a cost-effective way. Inventory piles up, expediting costs grow out of control, and fill rates decline, causing attention to be focused on the supply chain organization, which immediately points at the "poor forecasts" that come out of sales and marketing. The CEO gets excited, S&OP is hailed as the way to get demand and supply in balance, and the senior supply chain executive is tasked with putting this process in place. Where the disconnect often takes place, however, is with lack of engagement from the sales and marketing functions in the organization—the owners of customers and the drivers of demand. Nothing can make S&OP processes fail any faster than having sales and marketing be non-participants. In more than one company we've worked with, people describe S&OP as "&OP"—meaning that "Sales" is not involved.

Third, the very name "Sales and Operations Planning" carries with it a tactical aura. As argued in an upcoming section, many more functions besides Sales and Operations must be involved in order for effective business planning to take place. Without engagement from marketing, logistics, procurement, and particularly finance and senior leadership, these attempts at integrated business planning are doomed to being highly tactical and ultimately disappointing.

Thus, although the goals of S&OP are not incompatible with the goals of DSI, the execution of S&OP often falls short. Perhaps a new branding campaign is indeed needed, because in many companies, S&OP carries with it the baggage of failed implementations. Demand/Supply Integration is an alternative label and a new opportunity to achieve integrated, strategic business planning.

Signals that Demand and Supply Are Not Effectively Integrated

As our research team has worked with dozens of companies over the past 15 years, we have witnessed many instances where demand and supply are not effectively integrated. Commonly, our team is called in to diagnose problems with the forecasting and business planning processes at companies because some important performance metric—often inventory turns, carrying costs, expedited freight costs, or fill rates—have fallen below targeted levels. After we arrive onsite, we frequently hear about problems like those in the following list. Ask yourself whether any of these situations apply to your company:

- Does manufacturing complain that sales overstates demand forecasts, doesn't sell the product, and then the supply chain gets blamed for too much inventory?

- Does the sales team complain that manufacturing can't deliver on its production commitments and it's hurting sales?

- Does manufacturing complain that the sales team doesn't let them know when new product introductions should be scheduled, and then they complain about missed customer commitments?

- Does the sales team initiate promotional events to achieve end-of-quarter goals, but fail to coordinate those promotional activities with the supply chain?

- Does the business not take advantage of global supply capabilities to profitably satisfy regionally?

- Are raw material purchases out of alignment with either production needs or demand requirements?

- Does the business team adequately identify potential risks and opportunities well ahead of time? Are alternatives discussed and trade-offs analyzed? Are forward actions taken to reduce risk and meet goals, or are surprises the order of the day?

If these are common occurrences at your company, then the case might be that your Demand/Supply Integration processes might not be living up to their potential. The case might also be that one or more of the critical subprocesses that underlie the DSI "superprocess" might be suffering from inadequate design or poor execution.

The Ideal Picture of Demand Supply Integration

Figure 1-1 represents an "ideal state" of DSI. The circles represent functional areas of the firm, the rectangle represents the superprocess of DSI, the dark gray (dark purple in the e-book) arrows leading into the DSI process represent inputs to the process, and the lighter gray (red in e-book) arrows leading out of DSI represent outputs of the process.

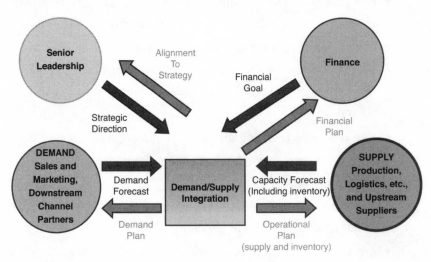

Figure 1-1 Demand/Supply Integration: the ideal state

It all begins with the two dark gray (dark purple) arrows labeled "Demand Forecast" and "Capacity Forecast." As future chapters clearly articulate, the demand forecast is the firm's best "guess" about what customer demand will consist of in future time periods. It should be emphasized that this is indeed a guess. Short of having a magic crystal ball, uncertainty exists around this estimate of future demand. Of course, the further into the future one estimates demand, the greater the uncertainty that exists. Similarly, the "Capacity Forecast" represents the best "guess" about what future supply capability will be. Just as is the case with the demand forecast, uncertainty surrounds any estimate of supply capability. Raw material or component part availability, labor availability, machine efficiency, and other supply chain variables introduce uncertainty into estimates of future capacity levels.

Let's begin with a simple example as a way of explaining how the DSI process needs to work. Assume that the "Demand" side of the business—typically sales and marketing, with possible input from channel partners—goes through an exercise in demand forecasting, and concludes that three months from the present date, demand will consist of 10,000 units of a particular product. Let's further assume that this demand forecast is reasonably accurate. (I know; that may be an unrealistic assumption, but let's assume it regardless!) Now, concurrently, the "Supply" side of the business—operations, logistics, procurement, along with input from suppliers—conducts a capacity forecast and concludes that three months from the present date, supply capability will consist of 7,500 units. Note that this outcome is far from atypical. *The fact is that demand and supply are usually* NOT *in balance.* So, more demand exists than supply. The question is, "*What should the firm do?*"

Several options exist:

- **Dampen demand.** This option could be achieved in a number of ways. For example, the forecasted level of demand assumes a certain price point, a certain level of advertising and promotional support, a certain number of salespeople who are selling the product with certain incentives to do so, a certain level of distribution, and so forth. Any of these demand drivers could be adjusted in an effort to bring demand into balance with supply. Thus, some combination of a price increase or a reduction in promotional activity could dampen demand to bring it in line with supply.

- **Increase capacity.** Just as the demand forecast carries with it certain assumptions, so, too, does the capacity forecast. Capacity could be increased through adding additional shifts, outsourcing production, acquiring additional sources of raw materials or components, speeding up throughput, and so forth.

- **Build inventory.** Often, the case is that in some months, capacity exceeds demand, whereas in other months, demand exceeds capacity. Rather than tweaking either demand or supply on a month-by-month basis, the firm could decide to allow some inventory to accumulate during excess capacity months, which would then be drawn down during excess demand months.

These are all worthy options for solving the "demand is greater than capacity" problem. The question, of course, is "Which one is the best for solving the short-term problem, while at the same time achieving a variety of other goals?"

The answer is, "It depends." It depends on the costs of each alternative and the strategic desirability of each alternative. Because each situation is unique, with different possible alternatives that carry with them different cost and strategic profiles, the need exists to put these available alternatives in front of knowledgeable decision-makers who can determine which is the best course of action. This is the purpose of Demand/Supply Integration, represented as the rectangle in Figure 1-1. The "Financial Goal" arrow represents the financial implications of each alternative, and the "Strategic Direction" arrow represents the strategic implications of each alternative. All these pieces of information from all these different sources—the Demand Forecast, the Capacity Forecast, the Financial Goal, and Strategic Direction—must be considered to make the best possible decisions about what to do when demand and supply are not in balance.

This simple example could be turned in the other direction. Suppose that the demand forecast for 3 months hence is 10,000 units, and that the capacity forecast for that same time period is 15,000 units. Now, the firm is faced with the mirror image of the first situation. Instead of dampening demand with price increases or reduced promotional support, the firm can increase demand with price reductions or additional promotional support. Instead of increasing production with additional shifts or outsourced manufacturing, the firm can reduce production with fewer shifts or taking capacity down for preventive maintenance. Instead of drawing down inventory, the firm can build inventory. Once again, the answer to the question of "What should we do?" is "It depends." Again, the correct answer is a complex consideration of costs and strategic implications of each alternative. The right people need to gather with the right information available to them to make the best possible decision—once again, DSI.

To further illustrate this "ideal state" of DSI, consider another example. This time, assume that the demand forecast for 3 months hence, and the capacity forecast for 3 months hence, are both 10,000 units (an unlikely scenario, but assume it anyway). Further, assume that if the firm sells those 10,000 units 3 months hence, the firm will come up short of its financial goals, and the investment community will hammer the stock. Now what? Now, both demand side and supply side levers must be pulled. Demand must be increased by changing the assumptions that underlie the demand forecast. Prices could be lowered, promotional activity could be accelerated, new distribution could be opened, and new salespeople could be hired. Which choice is optimal? Well, it depends. Simultaneously, supply must be increased to meet this increased demand. Extra shifts could be added, production could be outsourced, or throughput could be increased. Which choice is optimal? Well, it depends. The right people with the right information need to gather to consider the alternatives—again, DSI.

So far we have covered the inputs to the DSI process. An unconstrained forecast of actual demand is matched up against the forecasted capacity to deliver products or services. Within the Demand/Supply Integration process, meetings occur where decisions are made about how to bring demand and supply into balance, both in a tactical, short-term context *and* in a strategic, long-term context. Financial implications of the alternatives are provided from finance, and strategic direction is provided by senior leadership. However, Figure 1-1 also contains arrows that designate outputs from the DSI process. You should look at these outputs as *business plans*. Three categories of business plans result from the DSI process, as follow:

- **Demand plans** represent the decisions that emerge from the DSI process that will affect sales and marketing. If prices need to be adjusted to bring demand into balance with supply, then sales and marketing need to execute those price changes. If additional promotional activity needs to be undertaken to increase demand, then sales and marketing need to execute those promotions. If new product introductions need to be accelerated (or delayed), then those marching orders need to be delivered to the responsible parties in sales and marketing. The vignette from the beginning of the chapter represented a disconnect associated with communicating and executing these demand plans.

- **Operational plans** represent the decisions from the DSI process that will affect the supply chain. Examples of these operational plans are production schedules, inventory planning guidelines, signals to procurement that drive orders for raw materials and component parts, signals to transportation planning that drive orders for both inbound and outbound logistics requirements, and the dozens of other tactical and strategic activities that must be executed in order to deliver goods and services to customers.

- **Financial plans** represent signals back into the financial planning processes of the firm, based on anticipated revenue and cost figures that are agreed to in the DSI process. Whether the activity is financial reporting to the investment community or acquisition of working capital to finance ongoing operations, the financial arm of the enterprise has executable activities that are dependent upon the decisions made in the DSI process about how demand and supply will be balanced. Lastly, those signals come back to the senior leadership of the firm that the decisions that have been reached align with the strategic direction of the firm. These signals are typically delivered during the executive DSI meetings, which are corporate leadership sessions where senior leaders are briefed on both short- and long-term business projections.

Thus, in its ideal state, DSI is a business planning process that takes in information about demand in the marketplace, supply capabilities, financial goals, and the strategic direction of the firm, and *makes clear decisions about what to do in the future*.

DSI Across the Supply Chain

Up until now, we have talked about the need to integrate demand and supply within a single enterprise. In other words, how can insights about demand levels that might be housed in sales or marketing be shared with those who need to plan the supply chain? DSI processes are the answer. However, the ideal state of DSI doesn't need to be limited to information sharing within a single enterprise. Figure 1-2 represents a vision of how DSI can be expanded to encompass an entire supply chain.

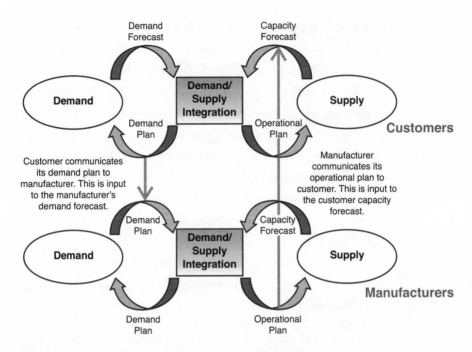

Figure 1-2 Demand/Supply Integration across the supply chain

Figure 1-2's representation of DSI is a simpler version of Figure 1-1 depicted in a simplified supply chain. The straight, vertical arrows show the possibilities for collaboration. First consider the arrow that leads from the customer's "Demand Plan" to the manufacturer's "Demand Forecast." Imagine, for example, that the "customer" in Figure 1-2 is a computer company, and the "manufacturer" is a company that produces microprocessors for the computer industry. The computer company's demand plan will include various promotional activities that it plans to execute in future time periods to take advantage of market opportunities. The manufacturer, the microprocessor company, would benefit from knowing about these promotional activities, because it could then be able to anticipate increases in demand from this customer. Such knowledge would be incorporated into the demand forecast for the microprocessor company.

Next consider the arrow that points from the manufacturer's "Operational Plan" to the customer's "Capacity Forecast." When the microprocessor company completes its DSI process, one output is an operational plan that articulates the quantity of a particular microprocessor that it intends to manufacture in future time periods. The customer, the computer company, would benefit from knowing this anticipated manufacturing quantity, particularly if it means that the microprocessor company will not be able to provide as much product as the computer company would like to have. Such a shortage would need to be a part of the computer company's capacity forecast, because this shortage will influence the results of the DSI process at the computer company. Thus, the outputs of the DSI processes at one level of the supply chain can, and should, become part of the inputs to the DSI process at other levels of the supply chain.

This same logic would apply if the "customer" were a retailer and the "manufacturer" were a company that sold its products through retail. The retailer's promotional activity, as articulated in the retailer's demand plan, is critical input to the manufacturer's demand forecast. Also, the manufacturer's projected build schedule, as articulated in the manufacturer's operational plan, is critical input to the retailer's capacity forecast. Companies use a variety of mechanisms to support this level of collaboration across the supply chain. These mechanisms can be as simple as a formal forecast being transmitted from the "customer" to the "manufacturer" on a regular basis. The mechanism can also be much more formalized, and conform to the Collaborative Planning, Forecasting, and Replenishment (CPFR) protocol as articulated by Voluntary Interindustry Commerce Solutions (VICS). Regardless of how this collaboration is executed, the potential exists for significant enhancements to overall supply chain effectiveness when DSI processes are implemented across multiple levels of the supply chain.

Typical DSI Aberrations

The "ideal state" of DSI as depicted in both Figures 1-1 and 1-2 are just that—ideal states. Unfortunately, a variety of forces often result in actual practice being far removed from ideal practice. I have observed a variety of "aberrations" to the ideal states articulated earlier, and three are so common that they are worth noting.

The first, and perhaps most insidious, of these DSI aberrations is depicted in Figure 1-3. The figure is simplified to highlight the aberration, which is known as *plan-driven forecasting*. Previously in this chapter, I discussed the not-uncommon scenario where forecasted demand fails to reach the financial goals of the firm. In the ideal state of DSI, that financial goal is one of the inputs to the DSI process, where decisions are made about how demand (and if necessary, supply) should be enhanced to achieve the financial goals of the firm. In a plan-driven forecasting environment, however, the financial goal does not lead to the DSI process. Rather, it leads to the demand forecast. In other words, rather than engaging in a productive discussion about how to enhance demand, the message is sent to the demand planners that the "right answer" is to simply raise the forecast so that it corresponds to the financial goal. This message can be simple and direct—"raise the forecast by 10%"—or it can be subtle—"the demand planners know that their forecast had better show that we make our goals"—but either way, the plan-driven forecasting aberration is insidious because it results in a forecasting process that loses its integrity. If downstream users of the forecast—those who are making marketplace, supply chain, financial, and strategic decisions—know that the forecast is simply a restatement of the financial goals of the firm, and not an effort to predict real demand from customers, then those users will stop using the forecast to drive their decisions. I have observed two outcomes from plan-driven forecasting:

- Supply chain planners go ahead and manufacture products that correspond to the artificially inflated forecast; the result is excess, and potentially obsolete, inventory.

- The supply chain planners say to themselves, "I know darn well that this forecast is a made-up number and doesn't represent reality. And since I own the inventory that will be generated by overproducing, I'm just going to ignore the forecast and do what I think makes sense." Here, the result is misalignment with the demand side of the company.

In both cases, plan-driven forecasting results in a culture where the process loses integrity, and forecast users stop believing what forecasters say.

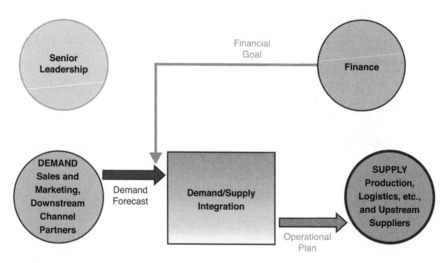

Figure 1-3 Typical DSI aberration: plan-driven forecasting

Figure 1-4 shows the second DSI aberration, what I call *DSI as a tactical process*. In this scenario, the responsibility of the demand side of the enterprise is to come up with a demand forecast, which is then "tossed over the transom" to the supply side of the enterprise, which then either makes its plans based on the forecast, or not. Often, no DSI process really is in place—no scheduled meetings where

demand-side representatives and supply-side representatives interact to discuss issues or constraints. When this aberration is in place,
significant risk exists for major disconnects between sales and marketing and supply chain. Without the information-sharing forum that
a robust DSI process provides, both sides of the enterprise usually
develop a sense of distrust, neither understanding nor appreciating
the constraints faced by the other. In addition to the siloed culture
that results from this aberration, the lack of engagement from either
senior leadership or finance makes this approach to DSI extremely
tactical. Oftentimes, the forecasting and planning horizons are very
short, and opportunities that might be available to grow the business
might be sacrificed, because demand and supply are not examined
from a strategic perspective. In this scenario, it is nearly impossible
for DSI to be a process that "runs the business." Instead, it is limited
to a process that "runs the supply chain," and engagement from sales
and marketing leadership becomes very challenging.

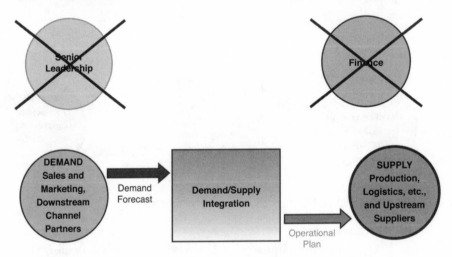

Figure 1-4 Typical DSI aberration: DSI as a tactical process

Figure 1-5 shows the final common aberration to the ideal state of
DSI: *lack of alignment with sales and marketing*. In this situation, little,
if any, communication gets back to the demand side of the enterprise

concerning the decisions made in the DSI process. This aberration is most problematic when capacity constraints are in force, resulting in product shortages or allocations. Recall the scenario described at the beginning of this chapter, where the sales organization at the apparel company was incentivized to sign up new accounts, while production problems were affecting deliveries to the firm's largest, most important current customer. As previously illustrated, the discussions that occur in the DSI process often revolve around what actions should be taken when demand is greater than supply. However, if no effective feedback loop exists that communicates these decisions back to the sales and marketing teams, then execution is not aligned with strategy, and bad outcomes often occur.

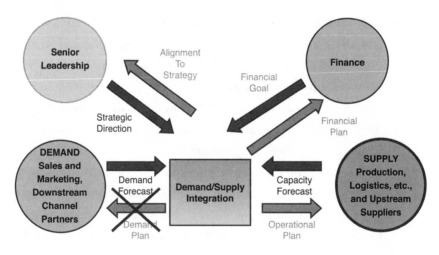

Figure 1-5 Typical DSI aberration: lack of alignment with sales and marketing

The aberrations described here are examples of typical problems faced by companies when their DSI processes are not executed properly. Aberrations like these often exist, even when the formal process design is one where these aberrations would be avoided. However, siloed cultures, misaligned reward systems, lack of training, and inadequate information systems can all conspire to undermine these process designs. The reader is encouraged to look carefully at Figure 1-1, which represents the ideal state of DSI, and carefully think through each of the input and output arrows shown in the figure. Wherever an arrow is missing, or pointed at the wrong place, an aberration occurs. Identifying gaps in the process is the first step to process improvement.

DSI Principles

Now that the ideal structure of DSI has been described, along with typical aberrations to that ideal, discussing some of the guiding principles that should drive the implementation of DSI at any company is appropriate. Three guiding principles are that

- **DSI should be demand driven.** Many years of supply chain research conclude that the most successful and effective supply chains are demand driven. In other words, supply chains are most effective when they begin with the voice of the customer. DSI processes should reflect this demand-driven orientation. The "Demand Forecast" arrow in Figure 1-1 represents this principle. The demand forecast is the voice of the customer in the DSI process. However, in too many instances, this customer voice is not as well represented as it should be, because sales and marketing are not as committed to or engaged in the process as they need to be. Because of culture, driven by measurement and reward systems, the weak link in many DSI implementations is the engagement from sales and marketing. Chapters 4, "Qualitative Forecasting Techniques" and 5, "Incorporating Market Intelligence into the Forecast," explore this phenomenon in greater detail.

- **DSI should be collaborative.** Figure 1-1 indicates that inputs to the process come from a variety of sources, both internal and external: sales, marketing, operations, logistics, purchasing, finance, and senior leadership represent the typical internal sources of information, and important customers and key suppliers represent the typical external sources of information. For this information to be made available to the process, a culture of collaboration must be in place. This culture of collaboration is one where each individual who participates in the process is committed to providing useful, accurate information, rather than pursuing individual agendas by withholding or misconstruing information. Establishing such a culture of collaboration is often the most challenging aspect of implementing an effective DSI process. Senior leadership must play an active role in developing and nurturing such a culture.

- **DSI should be disciplined.** When I teach undergraduate students, who often have little experience working in complex organizations, I often make the point that "what goes on in companies is *meetings*. You spend all your time either preparing for meetings, attending meetings, or doing the work that results from meetings." DSI is no different. The core of effective DSI processes are a series of meetings, and for DSI to be effective, the meetings that constitute the core of DSI must also be effective. This means discipline. Discipline comes in several forms. The right people must be in attendance at the meetings, so decisions about balancing demand and supply can be made by people who have the authority to make those decisions. Agendas must be set ahead of time and adhered to during the meetings. Discussion must focus on looking forward in time, rather than dwelling on "why we didn't make our numbers last month." To develop and maintain such discipline, an organizational structure must be in place where someone with adequate organizational "clout" owns the process and where senior leadership works with this process owner to drive process discipline. Once again, this points to organizational culture being critically important to DSI process effectiveness.

When these principles are embraced, then the "magic" of DSI can be realized, as shown in Figure 1-6.

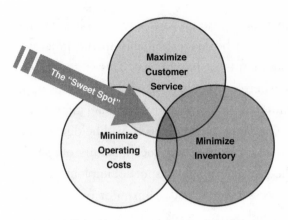

Figure 1-6 DSI "magic" comes from hitting the sweet spot.

The "sweet spot" shown in Figure 1-6 is the intersection of three conflicting organizational imperatives: maximizing customer service (having goods and services available to customers at the time and place those customers require), minimizing operating costs (efficient manufacturing processes, minimizing transportation costs, minimizing purchasing costs, and so on), and minimizing inventory. This "sweet spot" can be achieved, but it requires DSI implementation that is demand driven, collaborative, and disciplined.

Critical Components of DSI

This book is not intended to be a primer on the detailed implementation of DSI. Many other books do an outstanding job of providing that level of detail. However, in my experience of working with dozens of companies, I have observed that five critical components must be in place for DSI to work well. These components are

1. Portfolio and product review

2. Demand review

3. Supply review

4. Reconciliation review

5. Executive review

The following sections discuss each of these in terms of high-level objectives rather than tactical, implementation level details.

Portfolio and Product Review

The portfolio and product review is often absent from DSI processes, but including this step represents best practice. Its purpose is to serve as input to the demand review for any changes to the product portfolio. These changes typically come about from two sources: new product introductions and product (SKU) rationalization. New product forecasting is a worthy subject for an entire book,[1] and as such, I will not dwell on it here other than to say that predicting demand for new products, whether they are new-to-the-world products or simple upgrades to existing products, represents a complex set of forecasting challenges. Too often, new product development (NPD) efforts are inadequately connected to the DSI process, and the result is lack of alignment across the enterprise on the effect that new product introductions will have on the current product portfolio.

Another key element of the portfolio and product review stage of the DSI process is product, or SKU, rationalization. It is the rare company that has a formal, disciplined process in place for ongoing analysis of the product portfolio, and the result of this lack of discipline is the situation described by a senior supply chain executive at one company: "We're great at introducing new products, but terrible at killing old ones." This unwillingness to dispassionately analyze the overall product portfolio on an ongoing basis leads to SKU proliferation, and this leads to often unnecessary, and costly, supply chain complexity. By including product, or SKU rationalization as an ongoing element

---

[1] An excellent book-length primer on new product forecasting is *New Product Forecasting: An Applied Approach*, by Kenneth B. Kahn, M.E. Sharpe, 2006.

of the DSI process, companies have a strong foundation that permits the next step—the demand review—to accurately and effectively assess anticipated future demand across the entire product portfolio.

Demand Review

The demand review is, in essence, the *raison d'etre* for this book. The ultimate objective of the demand review is an unconstrained, consensus forecast of future demand. This meeting should be chaired by the business executive with P&L responsibility for the line of business, which constitutes the focus of the DSI process. This could be the company as a whole, or it could be an individual division or strategic business unit (SBU) of the company. Key, decision-capable representatives from the demand side of the enterprise should attend the demand review meeting, including product or brand marketing, sales, customer service, and key account management. One company that I've worked with has a protocol for its monthly demand review meeting. Sales and marketing personnel are invited and expected to attend and actively participate. Supply chain personnel are invited, but attendance is optional, and if they attend, they are not allowed to participate in the discussion! The intent is that the supply chain representatives are not allowed to chime in with statements like, "Well, we can't supply that level of demand." This protocol is one way that this company keeps the focus on unconstrained demand.

Because Chapter 8, "Bringing It Back to Demand/Supply Integration: Managing the Demand Review," covers the details of preparing for and conducting the demand review, I do not dwell on the specific agenda items or outcomes of the demand review here. For now, suffice it to say that the critical output of this stage in DSI is a consensus forecast of expected future demand.

Supply Review

The supply chain executive with relevant responsibility for the focal line of business being planned by the DSI process should chair the supply review meeting. The purpose of the supply review is to arrive at a capacity forecast, defined as the firm's best guess of its ability to supply products or services in some future time period, given a set of assumptions that are both internal and external. In addition, the supply review is that step in the DSI process where the demand forecast is matched up with the capacity forecast, and any gaps are identified, resolved, or deferred to future meetings.

The capacity forecast is determined by examining a number of pieces of information that are focused on the firm's supply chain. The components include supplier capabilities, actual manufacturing capabilities, and logistics capabilities. Supplier capabilities are typically provided by the purchasing, or procurement, side of the supply chain organization, and reflect any known future constraints that could result from raw material or component part availability. Manufacturing capabilities are determined through a number of pieces of information. These include:

- Historical manufacturing capacity
- Equipment plans, including new equipment that could increase throughput or scheduled maintenance that could temporarily reduce capacity
- Anticipated labor constraints, in terms of available specialty skills, vacation time, training time, and so on
- Improvement plans beyond equipment plans, such as process improvements that could increase throughput

Logistics capabilities can include any anticipated constraints on either inbound or outbound logistics, including possible transportation or warehousing disruptions. Altogether, these three categories of capabilities (supplier, manufacturing, and logistics) determine the overall capacity forecast for the firm.

After this capacity forecast is determined, it can be matched up against the demand forecast produced during the demand review stage of DSI. This is the critical point—where the firm identifies the kind of gaps described earlier in this chapter. This is also where DSI becomes the mechanism to plan the business. These gaps must be closed: when there is more demand than there is supply, or when there is more supply than there is demand. In some cases, these gaps are fairly straightforward, and solutions are fairly apparent. For example, if excess demand exists for Product A, and simultaneously excess capacity exists for Product B, simply shifting manufacturing capacity from B to A might be possible. The answer might be obvious. However, in other cases, as was described earlier in the chapter, the optimal solution to the demand-supply gap might not be so apparent. In those cases, one must take other perspectives into consideration, which is the reason behind the next step in the DSI process: the reconciliation review.

Reconciliation Review

If a firm were to stop the process with the supply review, described in the preceding section, it would have a perfectly serviceable, albeit tactical, S&OP process. The reconciliation review, along with the executive DSI review, transforms S&OP into DSI, for this is where the process is transformed from one that is designed to *plan the supply chain* into one that is designed to *plan the business*. The objective of the reconciliation review is to begin to engage the firm's senior leadership in applying both financial and strategic criteria to the question of how to balance demand with supply. The reconciliation review focuses on the financial implications of demand-supply balancing, and the meeting is thus typically chaired by the chief financial officer responsible for the line of business being planned. Attendees at this meeting include the demand-side executives (sales, marketing, and line-of-business leaders), and supply-side executives (the supply chain executive team), along with the CFO. Its aim is to have senior financial leadership lead the discussion that resolves any issues that emerged from the demand or supply reviews, and to ensure that all agreed-upon business plans are in alignment with overall firm objectives. At this point the discussions that have taken place in previous steps, which have typically focused on demand and supply of physical units, become "cashed-up." In other words, the financial implications of the various scenarios that have been discussed in the demand and supply reviews are now considered. Most unresolved issues can be settled at the reconciliation review, although some highly strategic issues might be deferred to the executive DSI review, discussed next.

Executive DSI Review

The executive DSI review is the final critical component of the DSI process, and it constitutes the regularly scheduled (usually monthly) gathering of the leadership team of the organization. The chair of this meeting is typically the CEO of the entity being planned, whether that is the entire firm or an identified division or SBU. The overall objective of the executive DSI meeting is to

- Review business performance
- Resolve any outstanding issues that could not be resolved at the reconciliation meeting
- Ensure alignment of all key business functions

This leadership meeting is where all key functions of the enterprise, from sales to marketing to supply chain to human resources to finance to senior leadership, come together to affirm the output of all the other pieces of the process, and where all functions can agree on the plans that need to be executed for the firm to achieve both its short- and long-term goals. In other words, this is where all functions gather to make sure that everyone is singing out of the same hymnal.

Clearly, for this process to be effective, the right preparation work must be completed before each of the scheduled meetings so that decision-makers have relevant information available to them to guide their decisions. Also, the right people must be present at each of these meetings. Both of these requirements lead directly to our next topic.

Characteristics of Successful DSI Implementations

An old quote, attributed to a variety of people from Sophie Tucker to Mae West to Gertrude Stein, says, "I've been rich and I've been poor, and rich is better." In a similar vein, I've seen good DSI implementations, and I've seen bad ones, and good is better! Based on the good, and bad, that I've seen, here are some characteristics of successful DSI implementations:

- **Implementation is led by the business unit executive.** In other words, *DSI cannot be a supply chain–led initiative if it is to be successful.* The most common reason for the failure of DSI is lack of engagement from the sales and marketing sides of an enterprise. Often, the impetus for DSI implementation comes from supply chain organizations, because they are often the "victims" of poor integration. Many of the 40+ forecasting and DSI audits that have been conducted by our research team have been initiated by supply chain executives, usually because their inventory levels have risen to unsatisfactory levels. The preliminary culprit is often poor forecasting, which is usually just the tip of the iceberg. Poor Demand/Supply Integration is usually to blame, and it is often due to lack of engagement from the sales and marketing sides of an organization. So DSI implementations absolutely need commitment of time, energy, and resources from sales and marketing, but the front-line personnel who need to do the work are often unconvinced that it should be part of their responsibility. This problem can be most directly overcome by having the DSI implementation be the responsibility of the overall business unit executive, who has responsibility for P&L, and to whom sales and marketing report.

• **Leadership, both top and middle management, is fully educated on DSI, and they believe in the benefits and commit to the process.** DSI is not just a process consisting of numerous steps and meetings. Rather, it is an organizational culture that values transparency, consensus, and a cross-functional orientation. This organizational culture is shaped and reinforced through the firm's leadership, and for DSI to be successful, all who are engaged in the process must believe that both middle and top management believe in this integrated approach. Such engagement from leadership is best achieved through education on the benefits of a DSI orientation.

• **Accountability for each of the process steps rests with top management.** Coordinators are identified and accountable for each step. Obviously, business unit and senior management have other things to do than manage DSI process implementation, so coordinators must be identified who have operational control and accountability for each step. DSI champions must be in place in each business unit, and this should not be a part-time responsibility. Our research has clearly shown that continuous process improvement, as well as operational excellence, requires the presence of a DSI leader who has sufficient organizational "clout" to acquire the human, technical, and cultural resources needed to make DSI work.

- **DSI is acknowledged as the process used to run the business, *not* just run the supply chain.** The best strategy for driving this thought process is to win over the finance organization, as well as the CEO, to the benefits of DSI. Without CEO and CFO engagement, DSI can easily be perceived throughout the organization as "supply chain planning." A huge benefit from establishing DSI as the way the business is run is that it engages sales and marketing. I have observed corporate cultures where DSI is marginalized by sales and marketing as "just supply chain planning, and I don't need to get involved in supply chain planning. That's supply chain's job." However, if DSI is positioned as "the way we run the business," then sales and marketing are much more likely to get fully engaged.

- **Recognition exists that organizational culture change must be addressed for a DSI implementation to be successful.** Several "levers" of culture change must be pulled for DSI to work:

  - *Values.* Everyone involved in the process must embrace the values of transparency, consensus, and cross-functional integration.

  - *Information and systems.* Although IT tools are never the "silver bullet" that can fix cross-functional integration problems, having clean data and common IT platforms can serve as facilitators of culture change.

  - *Business processes.* Having a standardized set of steps that underlie the DSI process is a key to culture change. Without clearly defined business processes, important elements of DSI can be neglected, leading to confusion and lack of integration.

- *Organizational structure.* Having the right people working in the right organizational structure, with appropriate reporting relationships and accountabilities, is a key facilitator to culture change.

- *Metrics.* People do that for which they are rewarded. What gets measured gets rewarded, and what gets rewarded gets done. These are clearly principles of human behavior that are relevant to driving organizational culture.

- *Competencies.* Having the right people in place to do the work, and providing the training needed for them to work effectively, are critical elements to successful DSI implementations.

DSI Summary

The remainder of this book focuses on how best to forecast demand. However, to bring the discussion full circle, the reader is referred to Figure 1-1—Demand/Supply Integration, The Ideal State. This book focuses on the arrow labeled "Demand Forecast." I strongly believe that without an effective Demand/Supply Integration process, an accurate demand forecast is not worth the paper it's written on (or the disk space it takes up on a computer system). I also strongly believe that this one arrow represents the critical beginning of the process. World-class DSI processes require world-class demand forecasting, and it is that to which the discussion turns.

2

Demand Forecasting as a Management Process

Our research team published an article several years ago titled, "The Seven Keys to Better Forecasting." Key number 1 is "Understand what forecasting is and what it is not," which relates directly to the topic of this chapter. We made the point that for too many companies, the management of forecasting begins and ends with the selection of forecasting software. For other companies, forecasting is the same as planning, and for still others, forecasting is an exercise in goal-setting. As emphasized in the article, *forecasting is a management process*, and like any other management process, it must be carefully organized, with attention paid to the people, processes, and tools that constitute forecasting management. This chapter turns your attention to appropriate ways to manage this critical business function. Specifically, the topics covered include:

- Defining forecasting, and how it is different from planning and goal-setting
- Defining key forecasting terms
- How forecasts are used by different functions in the firm
- The forecasting hierarchy
- Managing the forecasting process
- Overview of demand forecasting systems

- The role of forecasting techniques, both qualitative and quantitative
- The critical need to measure performance

This chapter explains some of these topics in detail. Other topics, such as forecasting techniques and the need to measure performance, are covered in more detail in chapters dedicated to these topics.

What Is Demand Forecasting?

Let's begin this section with a straightforward definition of demand forecasting:

A *demand forecast* is a firm's best estimate of what demand will be in the future, given a set of assumptions.

This simple definition leads to a variety of necessary elaborations. For one thing, a forecast is a "firm's best estimate" of future demand. In other words, it's a guess. It should, of course, be an educated guess, but it is nevertheless a guess. One critically important point that I often make, especially to executives, is that one thing upon which you can always depend is that your forecast will be *wrong*. Actually, two possibilities exist—your forecast will either be wrong, or you will be extremely lucky!

Also, this "best estimate of future demand" has, underneath it, a whole set of assumptions that are either explicitly stated (which is good) or implicitly assumed (which can be bad). These assumptions are both internal and external to the firm. Examples of internal assumptions are those firm activities that are typically put into place to generate demand, such as advertising expenditures, promotional activities, the opening of new distribution channels, the hiring of additional salespeople, and so forth. Another internal assumption is price levels. The economists teach us that demand curves are (usually) downward sloping, meaning that when prices fall, demand increases, and when prices rise, demand decreases. So an explicitly stated assumption about future pricing levels is critical to a future estimate of demand. Examples of external assumptions include expectations about future economic conditions, such as interest rates, unemployment levels, inflationary levels, behavior of the stock market, and so forth. Other relevant external assumptions involve competitive activity. If any information is available about key competitors and their new product plans, pricing strategies, promotional activities, and so forth, then that information will affect demand for the firm's products or services. A general rule is that the more explicitly stated these internal and external assumptions are, the better. Explicitly stated assumptions allow for a greater understanding of why the demand forecast was wrong—and remember: Forecasts are always wrong! Complete understanding of which assumptions held, and which did not, will aid tremendously in turning the forecasting process into one of continuous improvement.

The definition of demand forecasting can also be made more relevant by contrasting the process of demand forecasting with two other business activities: business planning and goal setting. Many companies that have participated in our research have problems with maintaining a healthy distinction between forecasting, planning, and goal setting. To contrast demand forecasting with business planning, the reader should refer to Figure 1-1 in Chapter 1. *Forecasts* are guesses about what might transpire in the future, both in terms of demand and supply. These forecasts serve as input to the DSI process. *Plans*, on the other hand, are *decisions made about what to do*, and they take the forms of various demand plans, operational plans, and financial plans. Think of these plans as "marching orders," or tasks that can be executed in an effort to capture the demand identified in the forecast. In an ideal DSI process, *forecasts precede plans*. In other words, decisions about what to do (plans) should be based on realistic assessments of future opportunities (forecasts). So a forecast is *not* a plan.

Neither is a forecast a *goal*, which is *an outcome that an individual or organization hopes to achieve*. Distinguishing between forecasts and goals is often quite problematic in business organizations. One business function that often has trouble distinguishing between goals and forecasts is sales. A quick story can illustrate this problem. At one company that participated in the audit research, I interviewed the vice president of sales and asked her about the way that salespeople completed their forecasts. Her response was, "Our people know that it would be suicide to forecast anything other than their targets." My follow-on question was, "But what if they don't think there is adequate demand from their customers to reach their targets?" Her reply: "You're not listening to me. They KNOW it would be suicide to forecast anything other than their targets." As Chapter 6, "Performance Measurement," discusses, the ultimate goal of a forecasting process is *accuracy*, or creating an estimate of future demand that is as near as possible to actual future demand. The preceding story illustrates a situation where the goal of the forecasting process for the salespeople in this company was by no means accuracy. Sales is not the only function in a company that confuses forecasting with goal setting. Marketing, in the form of brand management or product development, often confuses forecasting with goal setting as well. But the bottom-line principle is that a forecast is *not* a goal. A simple way to differentiate these two concepts is that a forecast is what we *think* will happen, whereas a goal is what we *hope* will happen.

The final elaboration for the definition of demand forecasting centers on the term *demand*. The reader should note that the term contained in the title of this book, and the terminology used throughout this book, is *demand forecasting*. A demand forecast is a best estimate of *demand*. Many other books and articles have been written on the subject of *sales forecasting*, and this term is generally accepted. It is, however, in my mind, the wrong term. The purpose of the process focused on in this book is to create the best possible estimate of future demand. So to formally define *demand*:

Demand is what customers would buy from us if they could.

An illustration can help to distinguish demand forecasts and sales forecasts. In spring of 2000, Sony Corporation introduced its PlayStation 2 (PS2) gaming console in Japan, followed by a fall 2000 launch in the United States and Europe. At the time of its introduction, Sony was faced with manufacturing delays that led to significant, and highly publicized, shortages on retailer shelves. Enthusiastic consumers did everything they could to secure one of the scarce PS2 consoles. I distinctly remember one of my undergraduate students in Fall 2000 who skipped class one day, and then reported to me at the next class period that he had waited all night in line to procure one of the scarce PS2s, which he then proceeded to sell on eBay the next day for more than double its list price. This is a well-publicized example of how *demand* and *sales* are often very different. Good reasons might exist for a company to produce a *sales forecast* (the best estimate of how many units will be sold in some future time period), including financial planning, especially reporting of revenue and profit projections to Wall Street, and short-term supply chain planning. However, a sales forecast is not the proper starting point for strategic business planning. If the company is capacity constrained and is able to sell every unit that it manufactures, then a sales forecast is likely to be 100% accurate! All the forecaster needs to do is to ask the question, "How many can we make?" The more useful, but more difficult, question is, "How many would customers buy from us if the product or service were available to them?" This is the demand forecast, which is the focus of this book.

Defining Some Key Terms

At this point, before delving deeper into forecasting as a management process, laying more foundation by clearly defining some important forecasting terms can be useful.

Forecasting Level

The *forecasting level* describes the level of granularity, or detail, in which a forecast is expressed. For example, consider a company such as Coca-Cola. One important forecasting question that Coca-Cola needs to answer is, "How many 12-pack, 12-ounce cans of Diet Coke will be demanded in the month of July in Knox County, Tennessee?" This is a Stock-Keeping-Unit by Location (SKUL) forecast. Another important question is, "How many 12-pack, 12-ounce cans of Diet Coke will be demanded in the month of July?" This is a stock-keeping unit (SKU) forecast. Another is, "How much Diet Coke, regardless of packaging size, will be demanded in the month of July?" This is a brand forecast. Still another is, "How much Coke, regardless of type (Coke Classic, Diet Coke, Coke Zero, Cherry Coke, and so on), will be demanded in July?" This is a product family forecast. As will be explained later in the section on the Forecasting Hierarchy, each of these forecasts is important for a different type of business planning, and each represents different forecasting levels. One general rule: Typically, the more granular the forecasting level, the less accurate the forecast. At lower levels of granularity, typically more random variation exists, which makes an accurate SKUL-level forecast much more difficult to achieve than an accurate brand forecast.

Forecasting Horizon

The *forecasting horizon* describes the length of time out into the future in which demand is being forecasted. In other words, if it is currently July, and you're forecasting demand that will occur in September, then the forecasting horizon is two months.

The question I am often asked is, "How far out into the future should a company forecast demand?" My answer has two parts. First is the question of the minimum length of the forecasting horizon. You should forecast demand at least as far into the future as constitutes the production lead time. In other words, if you are forecasting a product with a four-month lead time, it does your supply chain no good to have you create a forecast for what demand will be two months from now! The supply chain is unable to react to that two-month forecast, so it serves minimal purpose. So the minimum length of the forecasting horizon is the length of the lead time. Second is the maximum length of the forecasting horizon, and this is the amount of time it takes to create additional manufacturing capacity. Several years ago, our research team worked with a company that is in the business of manufacturing optical fiber. At that time, the estimated time to build a new optical fiber manufacturing facility was two years. Thus, this company's maximum forecasting horizon was two years. If it were to only forecast demand one year into the future, the information would not be sufficient to help it decide whether additional capacity needed to be built. So when the purpose of the demand forecast is supply chain planning, *the forecasting horizon should be no shorter than the production lead time, and at least as long as it takes to create new capacity.*

One observation is relevant to this discussion. Over the past 20 years or so, there has been a relentless move to offshore manufacturing to low-wage countries, particularly in Asia, but also in Latin America. Obviously, one consequence of this offshoring is to lengthen lead times for many manufactured goods. This trend impacts the demand forecasting process, because it requires demand forecasts to have a longer forecasting horizon. Another general rule is that the longer the forecasting horizon, the less accurate the forecast will usually be. A "guess" about what is likely to happen next month will be a better guess than a guess about what is likely to happen six months from now. Thus, one of the consequences of the trend to offshore manufacturing to low-wage countries is that the forecasts that are necessary to drive these longer lead times are usually less accurate than they would be if the manufacturing were taking place closer to the customer. As discussed in Chapter 1, when accuracy is lower, more inventory is needed to deliver acceptable service levels to customers.

Of course, other reasons exist for doing demand forecasting beyond supply chain planning. Financial planning and demand planning are also critical outcomes of a forecasting process (as managed by the DSI process). These planning horizons may be shorter or longer than the supply chain planning horizon. The principle in place here is that the forecast should be responsive to the needs of the user. The horizons for the firm's various planning tasks should drive the horizons for the demand forecast.

Forecasting Interval

The *forecasting interval* is the frequency at which the demand forecast is updated. For many manufacturing companies, the typical forecasting interval is monthly—a regular monthly "drumbeat" underlies the DSI, and thus the forecasting, processes. In some cases, however, the forecasting interval is considerably shorter. Some companies update their forecasts weekly, and some do it even daily! Typically, the dynamics behind such an accelerated forecasting interval is heavy promotional activity. Consumer packaged goods (CPG) companies often update forecasts very frequently, and these forecasts help to drive daily decisions about promotional activity that responds effectively to competitive dynamics, product availability, and so forth.

Forecasting Form

The *forecasting form* is the type of physical measurement in which the demand forecast is expressed. For example, Coca-Cola might express its forecast in cases, such as, "How many cases of Diet Coke will be demanded in July?" A chemical company might express its forecast in pounds. The manufacturer of optical fiber expressed its forecast in kilometers of optical fiber. In each of these examples, the forecast is expressed in a physical unit: cases, pounds, or kilometers.

One question I'm often asked pertains to the usefulness of having dollars (or euros or shekels or whatever kind of currency) as the forecasting form. In response, I typically pause, put a thoughtful look on my face, and respond with some wise remark like "Well, I think it's fine to forecast in dollars if your company manufacturers dollars (or euros or shekels)!" Apart from the dubious humor, I hope the point is clear. The demand forecasting process should begin with physical units, and a financial forecast—one expressed in dollars—should be arrived at by taking the physical forecast and applying pricing assumptions to it. I have worked with numerous companies that have a demand forecasting process that consists of asking their sales teams to submit forecasts for the upcoming quarter. Unfortunately, this forecast sometimes takes the form of, "My forecast for next quarter is $2.7 million." This is not a very useful way to run the supply chain, or the business, for that matter. You have no way of knowing whether this salesperson believes that he or she will sell one item valued at $2.7 million, 2.7 million items valued at $1.00, or something in between. Those individuals who are planning for operations, purchasing, logistics services, and inventory management obviously need a bit more detail to do their planning effectively. When sales teams are allowed to forecast in dollars, the result is usually not a forecast—what they believe will happen—but is rather a goal—what they hope will happen. Finance's responsibility is ultimately to "dollarize" the forecast; this exercise is a critical element of the DSI process. No way exists for business planners to identify gaps between forecast demand and the Annual Operating Plan (AOP), which is typically expressed in dollars, unless this dollarization exercise takes place. The resulting revenue, and profit, forecast will depend entirely on the mix of physical products or services that are forecasted. So once again, the rule is *begin with a forecast for physical units, and then dollarize that forecast to arrive at financial projections.*

How Forecasts Are Used by Different Functions in the Firm

Chapter 1 discusses how DSI is a process that requires a cross-functional orientation. When that DSI process is designed properly, it provides information back to a variety of organizational functions at the level, the horizon, the interval, and the form that is useful for each function. Table 2-1 provides a detailed summary of this concept. This table is not intended to be all inclusive, nor is it intended to be "true" for every organization. Rather, it illustrates how a robust forecasting and DSI process can provide useful forecasting outputs to each "customer" of the forecast in the organization.

The Forecasting Hierarchy

Closely related to the concept of the forecasting level discussed in the previous section is the concept of the *forecasting hierarchy*. The forecasting hierarchy is a way of illustrating the interconnections between different levels of forecasting granularity, and the relationships between those levels. The easiest way to conceptualize the forecasting hierarchy is as a pyramid with three "faces," as illustrated in Figure 2-1.

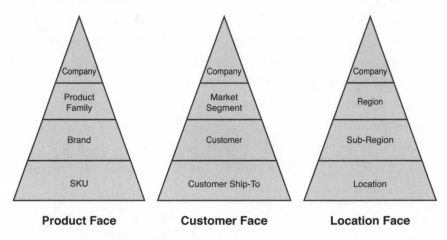

Figure 2-1 The forecasting hierarchy consists of three faces.

Table 2-1 Forecasting Needs of Different Functions

	Marketing	Sales	Finance	Sourcing	Logistics	Operations
Needs	Marketing needs a forecast of expected demand so it can effectively analyze promotional programs, new product introductions, and other demand-generation activities.	Sales needs a forecast of expected demand to be able to derive reasonable sales quotas, and to effectively allocate sales resources to those territories and customers where they can be best utilized.	Finance needs a forecast to be able to plan for working capital requirements and to be able to create financial projections for both Wall Street consumption and for government reporting.	*Strategic:* Long-term contracts with suppliers of needed raw materials, components, or capital equipment providers. *Tactical:* Manage short-term raw material and component part deliveries and inventories.	*Strategic:* Long-term contracts with transportation providers and warehousing assets. *Tactical:* Manage short-term transportation needs and day-to-day distribution center management.	Operations needs a forecast to be able to schedule manufacturing runs in the most efficient manner possible and to plan for expanded (or contracted) capacity to stay in line with market demand.
Level	Brand or product level forecast	Brand or product forecast by territory or customer	SKU forecast	Product forecast for strategic sourcing; SKU for tactical sourcing	Product forecast for strategic logistics; SKU by location for tactical logistics	SKU
Horizon	One to two quarters	One quarter to one year	One month to one year	Lead time to two years	Lead time to two years	Lead time to two years
Interval	Quarterly	Quarterly to annually	Monthly	Monthly	Monthly	Weekly to monthly
Form	Units that have been dollarized	Units that have been dollarized	Units	Units	Units	Units

For example, a series of product-oriented forecasts, customer-oriented forecasts, and location-oriented forecasts, each for a different purpose, is useful for Coca-Cola. The product-oriented forecasts could, at the lowest level of granularity, address the question of, "What will demand be for 2-liter bottles of Diet Coke in August?" Moving up the product hierarchy, the 2-liter bottle of Diet Coke forecast can be combined with the 12-pack can of Diet Coke forecast, and the 1-liter bottle of Diet Coke forecast, and every other way that Diet Coke is packaged, to answer the question, "How much Diet Coke will be demanded in August?" Moving up once again, the total Diet Coke forecast can be combined with the total Cherry Coke forecast, and the total Coke Zero forecast, and the total Classic Coke forecast to answer the question, "What will demand be for all Coke products in August?" Finally, the all-Coke forecast can be combined with the all-juice forecast and the all-water forecast, and any other product categories that Coca-Cola sells, to arrive at the overall company forecast. By the time the forecast reaches this highest level of the hierarchy, it will probably be expressed in dollars, because by that time a common unit measure might be meaningless, but the dollar forecast is useful for financial planning purposes.

The point of this example is that each forecast that is higher in the hierarchy is rolled up from forecasts at lower levels of granularity. The brand forecasts are rolled up from the SKU forecasts; the product family forecasts are rolled up from the brand forecasts; the company forecast is rolled up from the product family forecasts. The same logic is in place for the "customer face" of the hierarchy. The lowest level of granularity is particular customer ship-to locations (such as a particular Wal-Mart distribution center), which roll up to a forecast for an entire customer (Wal-Mart), which roll up to a forecast for a particular customer segment (mass merchandisers), which roll up to the company forecast as a whole. The "location face" of the hierarchy is conceptualized in the same way. The lowest level of granularity for a company might be a specific location (Tennessee), which rolls up to a subregion (United States), which rolls up to a region (the Americas), which rolls up to the company forecast as a whole. Each of these forecasts can be useful for specific demand planning, supply chain planning, or financial planning purposes. In most instances, these hierarchical relationships are defined within the data structures in place in the company's Enterprise Resource Planning (ERP) system. Particular SKUs are identified as being "children" of brand "parents," and brands are identified as being "children" of product family "parents," and so forth. A similar logic applies for the customer face and the location face of the hierarchy.

Why, you might ask, is this notion of a forecasting hierarchy important? It is important because if the forecasting hierarchy is structured appropriately, then executing the following forecasting rule becomes possible: *Put information into the forecast at the level you know it, and take information out at the level you need it.* Chapter 4, "Qualitative Forecasting Techniques," discusses the role that qualitative forecasting plays in the overall process, and Chapter 5, "Incorporating Market Intelligence into the Forecast," covers the incorporation of marketing intelligence into the forecast. In many cases, the qualitative judgments and market intelligence that come from sales and marketing, and that contribute so much richness to the forecast, come at higher levels of granularity than at the SKU level. For example, salespeople might not often discuss SKU-level details with their customers, but they might have extremely valuable information about changing demand patterns at the brand level or the product-family level. If the forecasting hierarchy is structured in an appropriate way, then those salespeople can contribute information at the level they know it (brand or product family), while at the same time, production, inventory, or transportation planners can extract information from the forecast at the level they need it (SKU or SKUL level).

Managing the Forecasting Process

A variety of factors determine what type of forecasting process will work best for any particular situation. In some companies, differences between business units result in different forecasting processes. Three overall areas must be examined prior to establishing a forecasting process. The process must accommodate differences between the natures of the customer base, the available data, and the products or services being forecasted, as discussed in the following sections.

The Nature of the Customer Base

Consider two forecasting situations, one with Boeing Corporation, the other with Hershey Foods. Both Boeing and Hershey Foods need to forecast demand for their products. In the case of Boeing Corporation, consider the commercial airframe portion of the company. A Google search of, "How many airlines are there in the world" led to an answer of approximately 300 commercial airlines. Assume that each of those customers buys airplanes, at least occasionally. Now consider Hershey Foods. Let's not even think about how many people consume products produced by Hershey Foods, but rather, focus on its retail customers. Another quick Google search revealed that just in the United States, 36,569 supermarkets exist. Add in the convenience stores (148,000), movie theaters, and other retail outlets that sell candy, and the number just in the United States is likely to be close to a half million. The point is that these extreme examples represent the different challenges faced by forecasters, depending on the nature of the customer base. In the case of Boeing, the best way to forecast demand for its commercial airplanes is to simply ask each and every one of its customers what their demand is likely to be; this would typically be done by the sales force. In the case of Hershey Foods, the best way to forecast demand for its several thousand SKUs from its several hundreds of thousands of customers is to perform sophisticated statistical analysis of historical demand patterns, and then apply qualitative adjustments to those statistical analyses. So the nature of the customer base impacts the decisions as to what processes are most effective for forecasting demand.

The Nature of the Available Data

Several categories of data are needed for effective forecasting; for example, historical demand data, data about customer promotional plans, data about past and future macro environmental trends, and competitor data. The availability, recency, and integrity of all these types of data can either enhance, or limit, the options available to forecasters.

One big problem that companies often face is the availability of historical *demand* data. Among the dozens of companies that have participated in our team's research efforts over the years, a great many lack good *demand* data. What is often used as the basis for analysis of historical demand is historical *shipments* rather than historical *demand.* A forecaster at a chemical company I worked with shared an example of this distinction with me. Imagine the following scenario. A customer calls into the customer service representative, and says, "I'd like to order 20,000 pounds of sodium benzoate [a preservative used in food processing], and I'd like it delivered in August." The customer service representative looks at the company's inventory and production plans, and responds, "Well, we can get you 15,000 pounds in August, but not 20,000. Could we deliver the remaining 5,000 pounds in September?" Perhaps the customer is flexible, and says "Okay, that will be fine." So that is what takes place: The order is placed, and the customer receives 15,000 pounds in August and 5,000 pounds in September. The question is this: Next year, when August demand is being forecasted for sodium benzoate, was last year's demand 20,000 pounds or 15,000 pounds? If the statistical analysis pulls from the *shipment* record, then the historical demand would be counted as 15,000 pounds in August and 5,000 pounds in September. However, if you recall the definition of *demand* (what customers would buy from a company if they could), then actual demand was 20,000 in August, which is when the customer preferred his shipment. In reality, this problem is easily solved. It is simply a matter of entering the order as 20,000 pounds in August, and then classifying the 5,000 pounds delivered in September as a back order. However, consider the following scenario, which changes the outcome. The customer calls and tries to order 20,000 pounds of sodium benzoate to be delivered in August. The customer service rep offers 15,000 pounds for August delivery, and the remaining 5,000 pounds for delivery in September. This time, the customer is not so accommodating, and says "No, I'll take your 15,000 pounds in August and I'll get the other 5,000 pounds from your competitor." In this situation, no order record exists that reflects the fact that a customer demanded—was ready to buy—20,000 pounds of sodium benzoate for August delivery. Next year, when statistical analysis is being done on historical demand, that historical demand will total only 15,000 pounds. This situation is much more difficult to solve. Companies can (and should!) create "lost order" records to account for such unfilled demand, but this is relatively rare. The moral of the story is that the quality of the demand data can inhibit forecasting excellence, and a process must be created to overcome this limitation.

Another example of how data availability can enhance, or limit, forecasting effectiveness lies in the efforts made to predict the impact of promotional activities on demand. As Chapter 3 discusses, this is a natural application of *regression analysis*, in which the forecaster is attempting to measure the impact that a particular variable (the independent variable) has historically had on overall demand for a product. A very common application of regression analysis is to use various types of promotional activities as the independent variable. However, if data that details these promotional activities is not available to the forecaster, then this type of analysis is not possible. Finally, the availability and quality of qualitative data, such as customer intentions, competitor actions, and macro environmental trends such as demographic or economic projections, might or might not be available to forecasters. In each of these examples, data availability, quality, and level of detail will influence the forecasting processes that are put into place.

The Nature of the Products

In the mid-1980s, I worked in sales for IBM, and I was part of a large team that supported the General Motors account. IBM sold many products and services to General Motors, but two products in particular, personal computers and mainframe computers, illustrate how the nature of the product influences the type of forecasting process that must be employed. In the case of personal computers, IBM sold millions of personal computers to tens of thousands of corporate customers and millions of individual consumers. The challenge was to predict the "run rate," or how many PCs would be demanded, by month, in future periods. Another challenge was to predict the "mix," or which of numerous different models of PCs would be demanded. In this situation, statistical analysis of historical demand, augmented by qualitative insights from the sales teams, was the logical process for forecasting these products. The other product, mainframe computers, were multi-million-dollar products typically purchased through a formal Request for Proposal (RFP) process, in which IBM would compete with other computer companies for the business. The challenge for forecasters was to predict the probability that IBM would win the business, so the preferred process for forecasting these products was heavy involvement from the sales teams. Such "project-based" businesses are very common, and such businesses face uncertainty in terms of the probability that the business will be "won," the mix of products that will ultimately be included in the awarded project, and the timing of when product delivery will be required by the customer. IBM faced challenges such as those faced by Boeing described earlier in this chapter, as well as other forecasting challenges such as those faced by Hershey Foods. The nature of the products determines the nature of the preferred forecasting process.

Another way that the nature of the product determines the nature of the process lies in the frequency of new product introductions. Two examples from companies that our research team worked with illustrate this phenomenon. Maxtor, which was purchased by Seagate in 2006, was a manufacturer of hard disk drives for the computer industry. Because of the screamingly rapid rate of technology innovation, the product life cycle for any of Maxtor's products was no more than about six months. New product introductions were nonstop, and were a combination of "new-to-the-world" products and upgrades to existing products, requiring complex "phase-in/phase-out" forecasting and planning. The other example is Amway, the direct sales company that sells everything from cosmetics to nutritional supplements to cleaning products. New product introduction is continuous and difficult to predict, particularly in its cosmetic business. All companies introduce new products, but the intensity of new product introductions, and the uncertainty surrounding them, differ from one company to the next, resulting in variation to the forecasting processes that are required to support these activities.

Another way that the nature of the product determines the nature of the process lies in the shelf life of the products being forecasted. For example, a company that our team worked with several years ago was in the business of manufacturing contact lenses. These products are built using highly capital-intensive manufacturing processes, and change-overs during the manufacturing process need to be minimized to keep these very expensive machines running at optimal throughput. The shelf life for these products is also long, and the storage costs are relatively low because the product itself is very small. These product characteristics led to a forecasting and planning process where the needed accuracy of the SKU forecasts is not particularly high. The company is far better off completing long manufacturing runs of individual SKUs, and then putting those products in inventory than they are trying to adjust manufacturing to be responsive to customer demand.

In sum, then, *the nature of the optimal forecasting process will depend on the nature of the business.* A forecasting process that works best for one company won't necessarily work best for a different company, because the nature of the customer base, the available data, or the products will be different from one company to the next. In fact, the nature of the optimal forecasting process for individual divisions in the same company might very well be different! A great example is Honeywell Corporation. Some of its businesses are build-to-stock in nature, where products are sold through retail to final consumers, and these businesses do their forecasting using sophisticated statistical modeling augmented by qualitative adjustments by the sales force. On the other hand, some of its businesses are project-based, where it competes for large multi-year contracts involving hundreds or thousands of different products. Clearly, the forecasting process that will work best in each of these situations will be different.

The Role of Forecasting Systems

In twenty-first-century industry, all business functions are supported by information systems that are designed to apply computer technology to make functions more effective. As one of many critical corporate business functions, demand forecasting is no different. Three critical functions are performed by demand forecasting computer systems:

- **Statistical engine.** Chapter 3 covers the value that comes from statistical analysis of historical demand. Statisticians have been working diligently for decades devising sophisticated algorithms that identify patterns in historical demand. Many of these algorithms are extremely labor intensive, and no human being should ever be expected to apply these algorithms by hand! Thankfully, computer programs have been developed to not only crunch the vast quantities of numbers that need to be crunched to apply these sophisticated algorithms, but to even choose the algorithm that best represents historical demand. Clearly, the great value that comes from application of statistical modeling is not possible without the "engine" that is embedded in demand forecasting software.

- **Data organizer.** Chapters 3, 4, 5, and 6 discuss the variety of data sources that must somehow be organized, coordinated, and made available to forecasting decision-makers. Here is a partial list of the various pieces of data that can prove to be very useful to forecasters:

  - Baseline statistical forecasts

  - Adjustments by marketing or product management

  - Adjustments by sales

  - Customer-provided forecasts

  - Historical accuracy and bias metrics

  One of the challenges faced by forecasters is to organize these various sources of data in such a way that they can be used effectively in the forecasting process. Information systems provide this organization and allow forecasters to access the right data to make the best possible decisions.

- **Data communicator.** In addition to organizing inputs to the forecasting process, information systems communicate the results of the forecasting process to the vast number of individuals and functions that use the forecast to do their planning. Table 2-1 listed just a few of the functions internal to the organization that need a forecast. Also, as discussed in Chapter 1, external users such as customers and suppliers need forecasts to be able to plan their businesses appropriately. Thus, integration with other corporate systems, and the ability to share information with upstream and downstream users, are critical components to a demand forecasting system.

Figure 2-2 presents a high-level overview of how demand forecasting systems should fit into an overall information technology architecture. Figure 2-2 offers two primary takeaways. One is that a professionally managed data warehouse is a critical component to the effectiveness of the demand forecasting process. One problem that can lead to a loss of credibility in the demand forecast is a lack of confidence in the integrity of the data being analyzed. The professionally managed data warehouse helps to maintain the integrity of that data. Another takeaway is that each of the systems that support the various corporate functions—logistics, production planning, finance, sales, marketing—should integrate "seamlessly" with the demand forecasting system. In other words, the less re-keying of data, or data translation from one system to another, the better. Chapter 7, "World-Class Demand Forecasting," covers this issue in more detail.

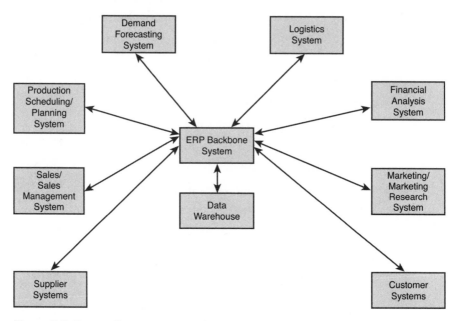

Figure 2-2 Forecasting system overview

Before leaving the topic of demand forecasting systems, I must emphatically make one critical point: *Systems are not silver bullets.* One phenomenon that I have observed in company after company is the tendency to try to fix poor forecasting with technology. For example, one company our research team worked with in the early 2000s manufactured consumer-oriented products sold in supermarkets and drug stores. Prior to calling our team in to help, it had invested in excess of $500,000 in a sophisticated forecasting system, and several months after the implementation of this system, the forecasting team was embarrassed to learn that their average forecast accuracy was considerably *worse* than it was before the system was implemented! They couldn't understand this result. However, our team discovered after interviewing many of the company's employees, that the real culprit was not an ineffective forecasting system, but rather a failure in change management. For example, salespeople we interviewed commented, "We've invested all this money in the new software, so that means I don't have to bother with forecasting anymore." Basically, many people who had previously contributed their insights about future demand now believed that the new forecasting software would do it all and be a "silver bullet." It wasn't. This example illustrates that the cultural values discussed in Chapter 1—specifically, the culture of collaboration and commitment to common goals—are a far more important contributor to demand forecasting excellence than the technology that is used to support the process. Unfortunately, many companies implement technology before a proper process is established, and without the change management strategy that is needed to create a collaborative culture. Clearly, technology support is important and the tasks described earlier—statistical engine, data organizer, and data communicator—are necessary for demand forecasting excellence. The technology does not solve the problem.

Forecasting Techniques

At the beginning of this chapter, I referenced an article my team had published called "The Seven Keys to Better Forecasting," and one of those keys is "know what a forecast is, and what it is not." The point was made that forecasting is a management process. What forecasting *is not* is an exercise in statistical analysis. Good forecasting does not simply mean, "Find the best statistical algorithm, and your problems will be solved." This section introduces the topic of forecasting techniques, which are discussed in far greater detail in Chapters 3 and 4.

Two categories of forecasting techniques exist: quantitative techniques and (not surprisingly) qualitative techniques. Quantitative techniques consist of the analysis of historical data in an effort to discover patterns of demand. Some patterns of historical demand can be identified because they repeat in a predictable way at certain times. For example, demand for pleasure boats is much higher in the spring and summer months than in the winter. Typically, *time series* techniques are appropriate statistical algorithms that can identify these patterns that repeat with time. Other patterns of historical demand can be identified because certain quantifiable variables have a predictable effect on demand. For example, Coca-Cola products tend to be quite responsive to promotional activity. When grocery stores put Coke on special, consumers buy more Coke. The amount of "lift" that can be expected from such promotional activity can be predicted using *regression* analysis. Chapter 3 covers some of these time series techniques as well as how regression analysis can be useful for demand forecasting. In both cases, time series analysis and regression analysis, forecasters use statistical techniques to try to identify patterns that occurred *in the past*. If such patterns exist, then those patterns can be projected into the future to arrive at a demand forecast.

It is critical to note, however, that these quantitative techniques are insufficient for excellence in demand forecasting. If you could be absolutely certain that the future will look exactly like the past, then you could stop with statistical analysis of historical demand. However, because that is so infrequently the case, qualitative techniques must be used to augment statistical analysis. Qualitative adjustments to these statistical techniques essentially answer the question, "How does the company think the future will look different from the past?" If you don't think the future will be different from the past, and if quantitative techniques can successfully find patterns, then those statistically generated forecasts provide as good a "guess" as you are likely to produce. However, forecasting excellence requires that the question be asked, "How is the future likely to look different from the past?" and such judgments can bring richness to the process as a whole.

Chapter 4 focuses on qualitative techniques. Some qualitative techniques, such as the Delphi method, are particularly useful for long-term, strategic forecasting. Others, such as sales force composite, are appropriate for both shorter-term, operational forecasting as well as strategic forecasting. Some qualitative techniques are particularly useful for forecasting new products, or for "phase in/phase out" situations where an old product is being replaced by a new version. Chapter 4 discusses each of these scenarios, citing pros and cons of each.

A key element of looking at forecasting as a management process is to recognize that in most cases, neither of these categories of techniques (quantitative or qualitative) is, by itself, sufficient for forecasting excellence. Figure 2-3 graphically illustrates the relationship between quantitative and qualitative forecasting, and the consensus process that brings these two different perspectives together. I refer to it as "the forecasting three-legged easel." Forecasting excellence is held up by the three legs of statistical forecasts, qualitative judgments, and the consensus process that brings the right people together to arrive at a forecast that everyone agrees is the best possible guess of future demand. If any of these three legs is missing, the easel falls over, and forecasting excellence tumbles down. See Chapter 3 for more about statistical techniques, Chapter 4 for more on qualitative judgments, and Chapter 8 for details on the consensus process.

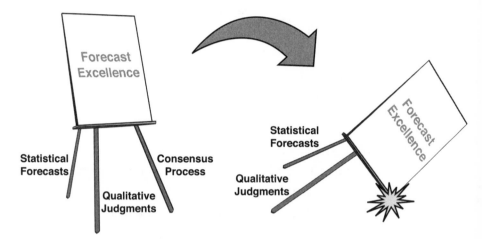

Figure 2-3 The forecasting three-legged easel

The Need to Measure Performance

An old management adage, possibly coined by Peter Drucker or W. Edwards Deming, says, "If you can't measure it, you can't manage it." Fortunately, forecasting is a management process that can indeed be measured, and that must be measured if it is to effectively contribute to an enterprise-wide demand-supply integration culture. Chapter 6 covers performance measurement in detail; as a preview, following are key bullet points from that upcoming discussion:

- Two dimensions of forecasting performance should be measured: accuracy and bias.

- Measuring accuracy has direct, functional results (such as serving as a surrogate for demand variability in safety stock inventory calculations), and indirect, motivational results (reflected in the phrase "what gets measured gets rewarded, and what gets rewarded gets done").

• No one buys or sells stock in a company because the company is good (or bad) at forecasting. Forecasting accuracy is a process metric. Process metrics are only interesting if they can be translated into outcome metrics, such as inventory turns, customer fill rates, reduced freight or raw material costs, or ultimately, profitability. Investment in forecasting excellence only makes sense if improved forecasting can lead to these other important outcome metrics.

So as is the case with any other management process, key performance indicators (KPIs) must be established to measure forecasting performance. Chapter 6 discusses, in detail, how to calculate those measurements and how to use them to enhance overall firm performance.

Summary

This chapter discusses a number of issues related to demand forecasting as a management process. Here are several key points:

• **Forecasting is a management process like any other.** A forecast is not a goal, nor is it a plan. It is also neither a piece of software, nor an exercise in statistics. It is a management process.

• **The nature of the business dictates the nature of the process.** All businesses are different in terms of their products, customers, available data, and people. Each forecasting process will have nuances that reflect these differences. However, you can find guiding principles throughout this book that are common to all forecasting processes.

- **Systems are not silver bullets.** Unfortunately, a company cannot buy forecasting excellence. It can buy tools to help facilitate forecasting excellence, but forecasting systems will not, by themselves, result in excellence.

- **Both qualitative and quantitative techniques are critical.** Statistical techniques are very useful for understanding what has happened in the past. Qualitative techniques are useful for predicting how the future will look different from the past. Both are critical.

- **What gets measured gets rewarded, and what gets rewarded gets done.** Although measuring forecasting performance can be a "pain in the neck," demand forecasting must be measured just as any other business process is measured.

With these process elements in mind, now turn your attention to the topic of statistical, or quantitative, forecasting.

3

Quantitative Forecasting Techniques

If you have picked up this book and immediately flipped to this chapter so that you can gain an in-depth understanding of statistical forecasting, including all formulas, assumptions, data requirements, and so forth, then you should put this book back on the shelf and find a different book. And there are lots of them. Plenty of other books will give you guidelines on techniques ranging from Box-Jenkins to Fourier Analysis to Spectral Analysis to Autoregressive Moving Average and so on. The books are excellent and the statistics are important elements of forecasting excellence. But that's not what this book provides.

This chapter focuses on the reasons behind statistical, or quantitative, forecasting techniques, and how managers should think about the role that statistical forecasting can, and should, play in the overall demand forecasting process. This chapter covers some of the more elementary statistical techniques that are often used by forecasters, and points out their pitfalls. It also touches upon some of the more sophisticated statistical modeling techniques and discusses how twenty-first century forecasting software helps the analyst choose the right model. The chapter concludes with a summary of the benefits that are gained from utilizing statistical analysis of historical demand, along with some cautionary words about over-reliance on statistical modeling.

The Role of Quantitative Forecasting

Quantitative forecasting is like looking in the rear-view mirror. The overall idea of statistical, or quantitative forecasting, is to look backwards, at history, to find and document patterns of demand. Chapter 2 presented the example of Hershey Foods. Demand planners at Hershey Foods can examine historical demand patterns and find important insights. One obvious insight they will observe is that in certain periods of the year, demand for Hershey chocolate products spikes. The weeks leading up to Halloween are high-demand periods, and the weeks immediately following Halloween are low-demand periods. Similar patterns occur around Valentine's Day (although the specific products or SKUs that are in high demand might be different at Valentine's Day than for Halloween). Similar patterns occur at Easter. Another type of pattern that demand planners at Hershey Foods might see is an overall upward (or downward) trend in demand for certain SKUs, brands, or product categories. Another pattern they might observe is a spike in demand in response to product promotions, or a dip in demand in response to competitive actions. Statistical analysis can help to not only identify these patterns, but also to predict the size and duration of the spike, or dip, in demand.

Statistical analysis can identify and predict two categories of patterns. The first category is those patterns that are associated with *time*. In the Hershey Foods example, the spikes that occur at Halloween or Easter are time-based patterns. Any overall upward (or downward) trends are also time-based. Identification and prediction of these time-based patterns are achieved through the use of various *time-series* statistical techniques. The second category of patterns is the influence that various factors other than time have on demand. An example of these "other factors" is promotional activity. If Hershey Foods embarks on an advertising campaign in the month before Halloween, then (hopefully) demand will increase as a result of that campaign. Demand planners need to know how much demand will change as a result of this advertising campaign. This type of question is best answered by *regression analysis*, which is a tool you can use to determine whether advertising campaigns—or other promotional activities—have influenced demand in the past, and if so, by how much. In both cases—time series and regression analysis—the demand planner is looking "in the rear view mirror" to find patterns that occurred historically. After those patterns are identified, they can be projected into the future, and *voilà!*—you have a forecast.

Time Series Analysis

Once again, time series techniques are a category of algorithms that are designed to identify patterns in historical demand that repeat with time. The three components of historical demand that these algorithms try to identify and predict are trend, seasonality, and noise.

- **Trend.** A trend is a continuing pattern of demand increase or decrease. A trend can be either a straight line (see Figure 3-1A), or a curve (see Figure 3-1B).

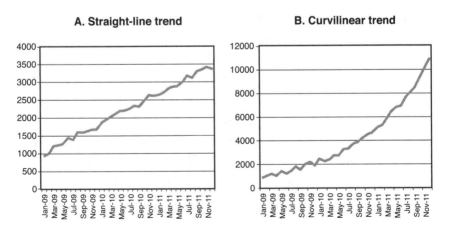

Figure 3-1 Trends in historical demand

- **Seasonality.** Seasonality is a repeating pattern of demand increases or decreases. Normally, we think of seasonality as occurring within a single year, and *cyclicality* as occurring over longer than a single year. Figure 3-2 shows a seasonal demand pattern.

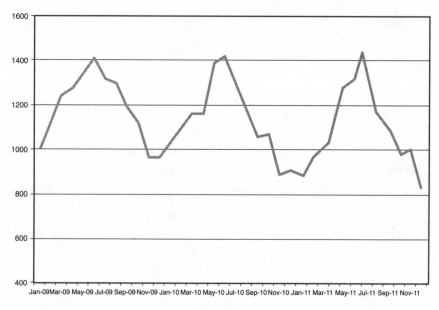

Figure 3-2 Seasonality in historical demand

- **Noise.** Noise represents random demand fluctuation. Noise is that part of the demand history that the other time series components (trend and seasonality) cannot identify. Figure 3-3 illustrates a demand pattern that contains no discernable trend or seasonality, but is simply noise. Most demand patterns contain some degree of random fluctuation—the less random the fluctuation (that is, the lower the noise level), the more "forecast-able" is the product or service.

One complicating factor, of course, is that it is often the case that all three of these components can be present in a stream of historical demand. The overall trend might be going up, while at the same time both repeating seasonal variation and random variation in the form of noise exist. The challenge, then, for forecasters who are examining these types of data patterns, is to find a time-series algorithm that can do the best possible job of identifying the patterns in the data, and then project those patterns into future time period. Let's begin with very simple techniques, and then move to more sophisticated ones.

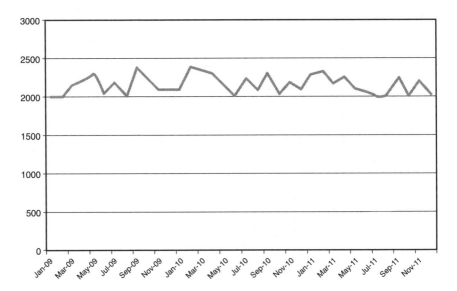

Figure 3-3 Noise in historical demand

Naïve Forecast

The simplest type of times series forecast is a *naïve forecast*. A naïve forecast is one where the analyst simply forecasts for future time periods whatever demand was in the most recent time period. In other words, if the forecaster is using a naïve forecasting approach, then the forecast for February, March, April, and all future months simply consists of whatever demand was in January. Then, when forecasting for March, April, May, and all future months, the forecast consists of whatever demand was in February. This obviously is a very simple procedure that fails to take into account any trend, seasonality, or noise that might be present in historical demand. Thus, other approaches are more widely, and effectively, used.

Average as a Time Series Technique

Except for a naïve forecast, the simplest form of time series analysis is a simple average. An average can be expressed arithmetically in the following formula:

$$Forecast_{t+1} = Average\,Demand = \sum_{t=1}^{N} \frac{D_t}{N}$$

where D = Demand and N = Number of periods of demand data.

In other words, when using a simple average as a forecasting technique, then next month's forecast, and every future month's forecast, is the average level of demand from all previous months.

One demand pattern exists in which a simple average is the best forecasting technique to use, and that is a pattern of random data, with neither a detectable pattern of trend or seasonality. Figure 3-4 illustrates how the average responds to this type of data stream. In this and all subsequent figures in this section, the "forecast" data point represents the average of all previous "demand" data points. For example, if the forecaster is working on her forecast in January 2011, the average of all previous demand data points is 2,209 units. Her forecast, then, for February 2011, March 2011, April 2011, and all subsequent periods in her forecasting horizon, would be 2,209 units. If this forecaster's current month were August 2011, then the average of all previous demand data points is 2,200 units, and all the forecasts in her upcoming forecast horizons would be 2,200 units.

When only noise is present, then "spikes" are offset by "dips," and the average demand from all previous periods is as good a forecast as the analyst can develop. Apart from this one fairly simple demand situation, though, using a simple average has significant pitfalls.

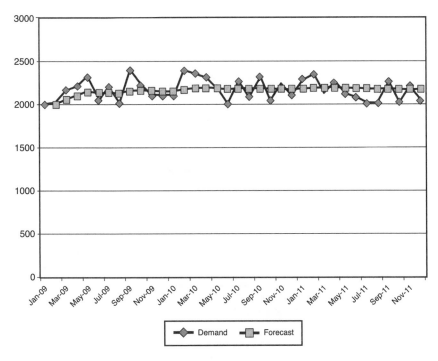

Figure 3-4 Average demand as a forecast, noise only

One demand pattern that does not lend itself well to using a simple average is a pattern where either an upward or downward trend is present. Figure 3-5 illustrates this demand pattern, and the pitfall from using an average. In this figure, each forecast moves further and further away from the demand trend line. The arithmetic reason that the forecast becomes worse and worse as each month goes by is that all previous months are used in the average calculation. For example, when the analyst is at November 2011, she is averaging all the previous demand from January 2009 through October 2011, and those early months of low demand keeps pulling the forecast further and further from the trend line.

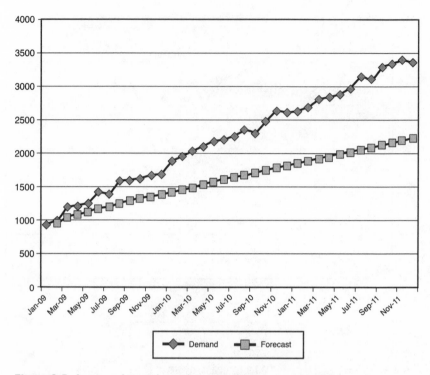

Figure 3-5 Average demand as a forecast, linear upward trend

Another common demand pattern that does not lend itself to being forecasted using simple average is a pattern of seasonality. Figure 3-6 illustrates the problems involved in this situation.

In this case, the forecast once again consists of the average of all previous demand data points. By the time the first peak is followed by the first trough, then each "high" is offset by a "low" and the forecast simply flattens out, in much the same way that it reacts to demand data that consists of only noise. Clearly, a simple average is not a good tool for modeling seasonal demand.

A final demand pattern that an average has a hard time modeling is a pattern that includes a change in the overall level. Consider the example found in Figure 3-7. You can see a situation where in January 2010, something big happened. Perhaps a competitor went out of business. Perhaps an entirely new market was opened. Whatever the reason, the base level of demand changed from somewhere in the neighborhood of 2,200 units to somewhere in the neighborhood of 3,100 units. When using a simple average, any forecast that is done following December 2010 will fall short of the new level. In fact, although the forecast will eventually get close to the new level, it will never get there and will asymptote to the new level. Why? The reason again lies in the use of irrelevant data. If a simple average is used, all the data from January through December 2009 is included in the average. Demand that occurred prior to the level change is not relevant, but keeping those data points in the calculation prevents the forecast from being correct.

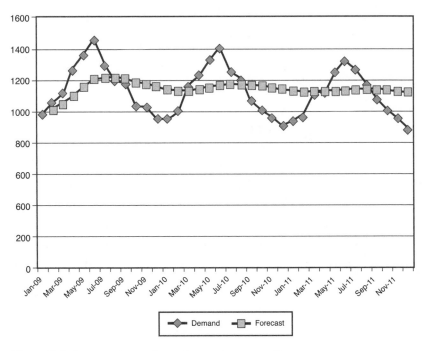

Figure 3-6 Average demand as a forecast, seasonality

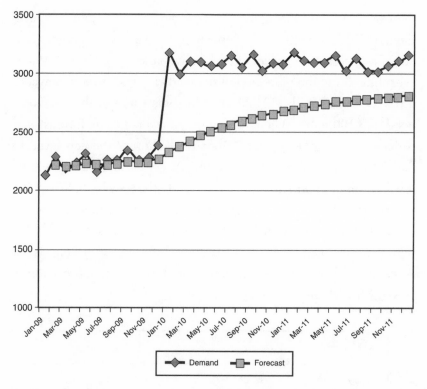

Figure 3-7 Average demand as a forecast, level change

Moving Average as a Time Series Technique

In two of the cases previously discussed, using a simple average to arrive at a forecast worked poorly because too much old, irrelevant data was used to calculate the average. In the case of the upward trend (refer to Figure 3-5), the forecast continues to be farther and farther away from the actual demand, and in the case of the level change (refer to Figure 3-7), the forecast never catches up to the new level, all because old, irrelevant data is included in the average calculation. This deficiency in the average can be overcome by using a *moving average*. A moving average is calculated using the following formula:

$$F_{t+1} = \frac{(D_t + D_{t-1} + D_{t-2} + \cdots + D_{t-[N-1]})}{N}$$

where F_{t+1} = Forecast for period t+1

D_{t-1} = Demand for period t–1

N = Number of periods in the moving average

For example, the equation for a three-period moving average is:

$$F_{t+1} = \frac{(D_t + D_{t-1} + D_{t-2})}{3}$$

Similarly, the equation for a four-period moving average is:

$$F_{t+1} = \frac{D_t + D_{t-1} + D_{t-2} + D_{t-3}}{4}$$

When using a moving average, the forecaster can decide how many periods are relevant, and thus eliminate irrelevant demand history from the calculation.

Figure 3-8 illustrates the effect of using three different moving average calculations on a demand history that contains an upward trend.

As the figure shows, the three-month moving average closely follows the actual demand trend line, although always falling somewhat underneath the actual demand. The six-month moving average also follows the demand trend, but falls a bit further under the actual trend, and the 12-month moving average follows the same pattern. Thus, in the case of a linear trend, either upward or downward, a relatively short-period moving average provides an excellent algorithm for forecasting demand.

The other demand pattern where a simple average is undermined by old, irrelevant data is the pattern with a level change (refer to Figure 3-7). With this demand pattern, a demand forecast that uses a simple average asymptotes to the new level, but never quite "catches up." Again, a moving average can provide a much more useful forecast in this case, as illustrated by Figure 3-9.

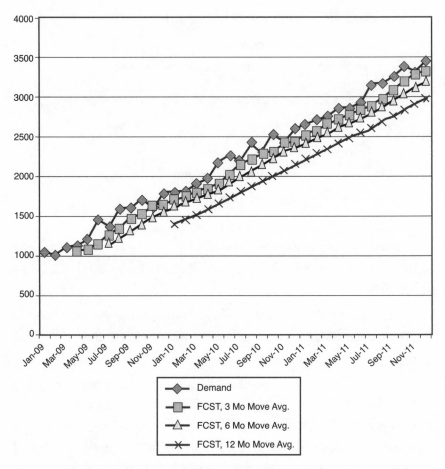

Figure 3-8 Moving average as a forecast, linear upward trend

In this scenario, the three-month moving average moves the forecast up to the new level quickly (in this case, in 3 months!). The 6-month and 12-month moving averages move the forecast up to the new level more slowly, but even the 12-month moving average gets the forecast up to the new level eventually. Clearly, then, the moving average methodology overcomes the pitfalls of a simple average approach in some demand patterns.

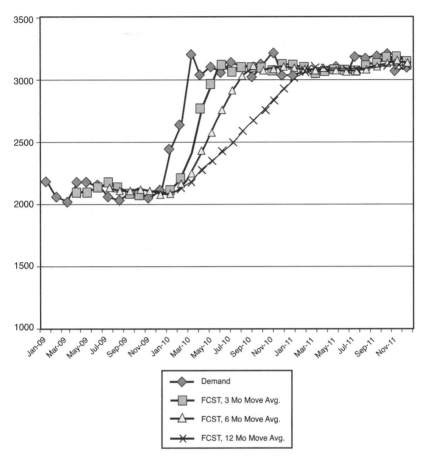

Figure 3-9 Moving average as a forecast, level change

However, in other demand patterns, a moving average is not the best tool for overcoming the deficiencies of a simple average. One such demand pattern is a seasonal pattern. As illustrated in Figure 3-10, a moving average of either 3, 6, or 12 months fails to provide adequate modeling of seasonal demand patterns. The 3-month moving average does respond to the highs and lows of seasonal demand, but the average lags behind the peaks and valleys by 3 months. In addition, the 3-month moving average flattens out the peaks and valleys. Similarly, the 6-month moving average lags by 6 months, and it flattens the peaks and valleys even more than the 3-month average. In the 12-month average, all the peaks offset the valleys, and the forecast looks remarkably like a simple average. Thus, a moving average is not a particularly useful tool for modeling demand when seasonal patterns are present. It should be pointed out, of course, that seasonal patterns are common, which makes a moving average less than ideal as a "go-to" forecasting methodology.

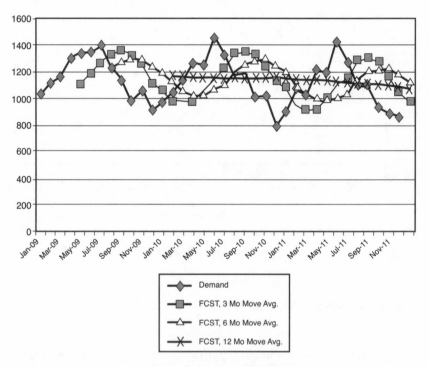

Figure 3-10 Moving average as a forecast, seasonal demand

Exponential Smoothing

As you saw in the preceding section, using a moving average is a good way to eliminate old, irrelevant data from the average calculation. It is, of course, a bit of a "blunt instrument" in achieving that end, in that it completely ignores any data that is older than the number of periods being considered. A different, and more sophisticated and flexible approach for determining how to use historical data is *exponential smoothing*. This approach allows the forecaster to decide how much weight should be applied to very recent data points, and how much should be applied to more distant data points. In moving average, it's all or nothing—either the previous data point is included in the calculation or not. In exponential smoothing, previous data points can be given a little or a lot of weight, at the discretion of the analyst.

The formula used to calculate a forecast using exponential smoothing is:

$$F_{t+1} = \alpha D_t + (1 - \alpha)F_t$$

where F_{t+1} = Forecast for period t+1

$0 < \alpha < 1$

D_t = Demand for period t

F_t = Forecast for period t

Although the introduction of a Greek character (α) might seem intimidating, you can think of it simply as a way to apply more or less weight to more recent observations. For example, when α is set at .1, then the most recent observation has a very small amount of extra weight attached to it, and for all practical purposes, all previous data points are weighted (more or less) the same. On the other hand, when α is set at .9, then considerable weight is assigned to the most recent data point, and much less weight is assigned to previous observations. What makes this technique exponential in nature is the fact that the second term in the equation—$(1 - \alpha)F_t$—involves the current period's *forecast*. The current period's forecast includes weighting from *previous* period's demand, and thus, the effect is that the model is exponential in nature.

To better understand the effect that different levels of α might have on a forecast, take a look at a few examples. Figure 3-11 shows a demand pattern that includes a change in level, and the effect of different levels of α is very clear. When α is set at .1, the result is that nearly equal weight is given to each of the historical demand points. Thus, the forecast looks very much like a simple average, with the problem being that the forecast never climbs to the new level, but rather asymptotes to it after considerable time getting there. At the other extreme, when α is set to .8, then the most recent observations are weighted very heavily, and the forecast very quickly rises to the new demand level. A general rule, then, is that α should be set high when there is a level change.

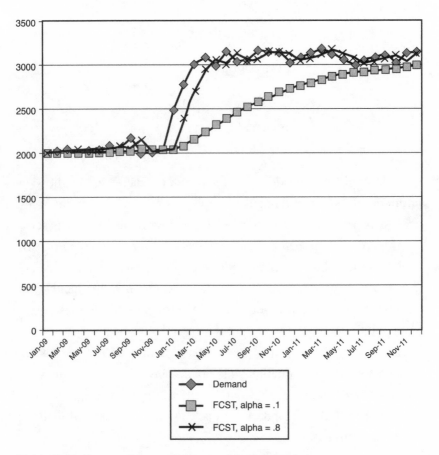

Figure 3-11 Exponential smoothing as a forecast, level change

Figure 3-12 shows a different example, where there is neither trend, level change, nor seasonality, but rather just noise. Here, when

α is set high, such as .8 from the example, then the fact that consider-able weight is placed on the most recent data point leads the forecast to react very quickly to the noise. In essence, the forecast is "chasing the noise." At a low level of α, such as the .1 case in the example, then the forecast looks very much like a simple average. As Figure 3-4 showed earlier, a simple average is probably as good a forecast as can be put together when all there is in demand history is noise. Thus, another general rule is that the noisier the demand history, the lower α should be.

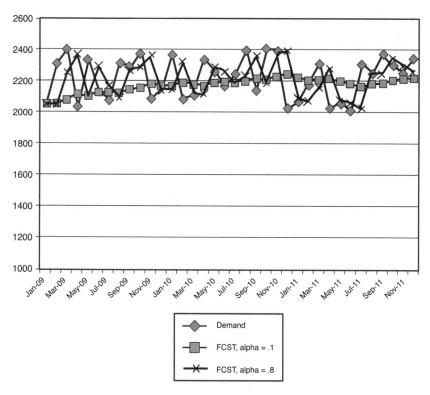

Figure 3-12 Exponential smoothing as a forecast, noise only

A final example shows how exponential smoothing reacts to seasonal demand data. Figure 3-13 shows a demand pattern that is clearly seasonal. When α is set to .1, the forecast reacts in a similar way to when we used simple average. The demand peaks offset the demand troughs, and the forecast becomes (more or less) a straight line. However, when α is set to .8, there is rapid reaction to the seasonal pattern. The forecast follows the seasonal pattern quite closely, but is lagged by a couple months, and so misses the timing of the peaks and troughs.

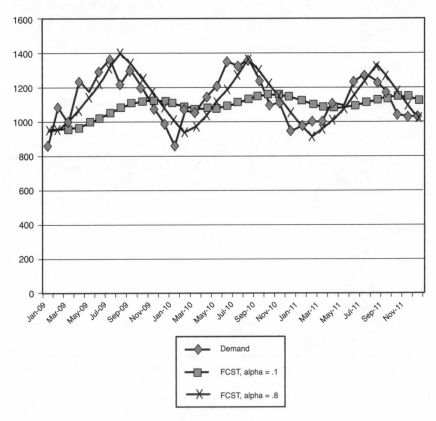

Figure 3-13 Exponential smoothing as a forecast, seasonal demand

Thus, exponential smoothing overcomes some of the pitfalls of both simple averages and moving averages, and because α can be set to any value between 0 and 1, the analyst has considerable flexibility to adjust the model to fit the data. How is the level of α best determined? A variety of *adaptive smoothing* techniques can use percent error calculations to help you hone in on the right level of α. Also, other complications exist that involve data patterns that need even more flexibility. For example, sometimes a trend exists in the demand data. An algorithm called *exponential smoothing with trend* would now be appropriate, where in addition to the smoothing constant (α), a trend constant (β) is introduced. In other cases, the data has both trend and seasonality. In these cases, *exponential smoothing with trend and seasonality* might be helpful, because α and β are now joined by γ, the seasonality constant. You can use adaptive techniques to transform these algorithms into *adaptive exponential smoothing with trend* and *adaptive exponential smoothing with trend and seasonality*. Here, I refer the reader to the first paragraph in this chapter—my intention is not to provide a comprehensive catalog, nor a detailed statistical explanation, behind all these computationally complex and statistically sophisticated methods for modeling historical demand. Other books are already in print that can provide this background.

Now that these various time series approaches have been discussed, a couple of questions remain for the working demand planner. The first question is, "How in the world do I decide which of these highly complex and sophisticated techniques to use?" Thankfully, the answer to this question is that if the analyst has a twenty-first century statistical forecasting software system in place, then he or she doesn't have to decide—the system will decide! Chapter 2 discussed the primary functions of a demand forecasting system, and one of those functions was what I described as a forecasting "engine." In twenty-first century forecasting systems, this engine is undoubtedly "expert" in nature. What the forecasting system is expert in is the selection of the most appropriate algorithm that best characterizes the historical demand data made available to it. Whether the forecast is being done at the SKU, brand, or product family level, the sequence of steps that an expert system will follow to create a forecast is

1. Access the demand history. Hopefully, that demand history represents true demand, and not just sales. Also hopefully, adequate historical demand exists to allow various statistical algorithms to identify patterns. Various algorithms require different numbers of historical data points to be able to estimate their parameters. Finally, and also hopefully, these data reside in a professionally managed data warehouse that is updated regularly. A statistical forecast is of little value if it is not created based on accurate, credible data.

2. From its catalog of statistical algorithms, the system applies the first time series methodology to the demand data accessed in step 1.

3. The system calculates the forecast error that would have been generated had that methodology been used, and stores this forecast error. Chapter 6, "Performance Measurement," discusses the term *forecast error* in great detail. For present purposes, think of forecast error simply as the difference between forecasted demand and actual demand.

4. The system then applies the second time series methodology to the demand data. It again calculates the forecast error that would have been generated had this second methodology been used. It compares the forecast error to the error from the first methodology, and whichever methodology is lower "wins," and remains stored.

5. The system then goes on to the third methodology and repeats the process. Some methodologies, particularly variations on exponential smoothing, require the estimation of various parameters, such as α, β, and γ, and different adaptive techniques will be applied to arrive at the best possible forecast for that particular methodology. This sequence continues through all the various time series methodologies that are included in

the software system. After all the methodologies have been tried, the system arrives at the one that would have generated the lowest error.

6. The system then uses the selected methodology to project into the future for the required forecasting horizon.

Although expert systems that follow the preceding sequence are wonderful time savers and add considerable power to the demand planner's arsenal, they must be utilized cautiously to avoid the phenomenon sometimes referred to as *black-box forecasting*. Black-box forecasting occurs when the analyst pours numbers into an expert forecasting system, and then takes the "answer" recommended by the system without questioning its reasonableness. An example can illustrate the danger that can come from black-box forecasting. Several years ago, our research team performed a forecasting audit for a company that was in the business of manufacturing and marketing vitamins and herbal supplements. These products were sold through retail, primarily at large grocery chains, drug chains, and mass merchandisers. Demand was variable, seasonal, and promotion driven, but reasonably forecastable. Then, an interesting event took place. In June of 1997, the popular ABC news magazine *20/20* aired a story about an herbal product called St. John's Wort, which had been touted in Europe as an herbal alternative to prescription anti-depressants. The *20/20* report was very complimentary of St. John's Wort, and presented it in a very positive light. Figure 3-14 shows what happened to consumer demand for St. John's Wort. (The numbers are made up, but the effect is consistent with actual events.) Demand skyrocketed. Demand remained high for a period, then gradually, over a period of a year or so, returned back to the level it had previously been before the broadcast. Now, imagine yourself as a forecaster for St. John's Wort in April of 1999. If the demand history that's shown in Figure 3-14 were to be loaded into an expert forecasting system, a pattern would undoubtedly be identified. The time-series analysis would project a dramatic jump in demand, followed by a slow decline. But unless the story about St. John's Wort were rebroadcasted on *20/20*, then such a forecast would be grossly high. The point, then, of this example is twofold. First, the analyst must understand the dynamics behind historical demand, and not simply rely on an expert forecasting system to do the job completely. Second, just because something happened in the past doesn't mean that it will happen again in the future. This means that the insights that come from understanding historical demand patterns through time-series analysis must be augmented by insights that come from the judgments of people. Chapter 4, "Qualitative Forecasting Techniques," returns to this point in great detail.

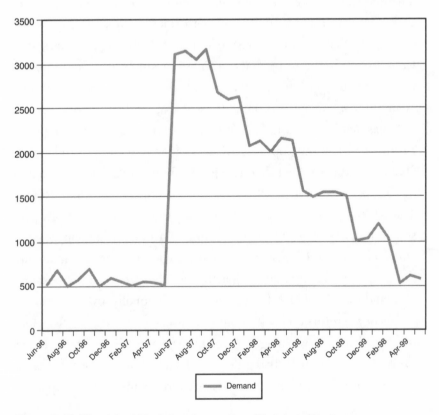

Figure 3-14 The risk of black-box forecasting: St. John's Wort

The second question that the working demand planner must face is, "If I have 10,000 SKUs in my product portfolio, how in the world do I manage that?" The answer to this question is a bit more complicated, and requires a revisiting of the forecasting hierarchy discussed in Chapter 2, "Demand Forecasting as a Management Process." Although forecasting systems can certainly handle the amount of data involved in doing 10,000 SKU-level forecasts, the SKU level might not be the most appropriate level at which to forecast. The following example might prove instructive. Suppose you have two SKUs you want to forecast. Table 3-1 shows the previous 12 months of historical demand, and Figure 3-15 shows a scatter plot of this historical demand.

Table 3-1 Example of SKU-Level Forecasting

	SKU #1	SKU #2
Jan-11	121	79
Feb-11	126	174
Mar-11	204	196
Apr-11	298	202
May-11	328	272
Jun-11	479	221
Jul-11	777	23
Aug-11	686	214
Sep-11	329	671
Oct-11	593	507
Nov-11	241	959
Dec-11	1181	119

What you see in this scatter plot is historical demand that appears largely random. Not much hope exists of finding a statistical algorithm that will effectively identify the pattern of either SKU. However, if you combine these two SKUs into a single product family, the picture looks very different. Table 3-2 and Figure 3-16 show both the preceding table and scatter plot, but re-created with the addition of the product family to the mix.

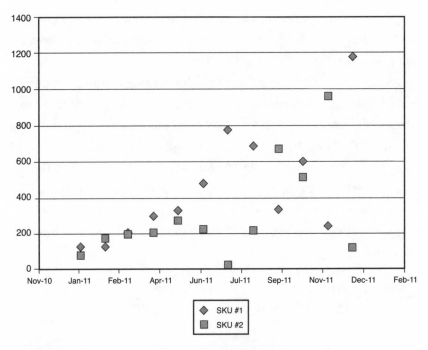

Figure 3-15 SKU level forecasting example: scatter plot

Table 3-2 Example of Product Family Level Forecasting

	SKU #1	SKU #2	Product Family
Jan-11	121	79	200
Feb-11	126	174	300
Mar-11	204	196	400
Apr-11	298	202	500
May-11	328	272	600
Jun-11	479	221	700
Jul-11	777	23	800
Aug-11	686	214	900
Sep-11	329	671	1000
Oct-11	593	507	1100
Nov-11	241	959	1200
Dec-11	1181	119	1300

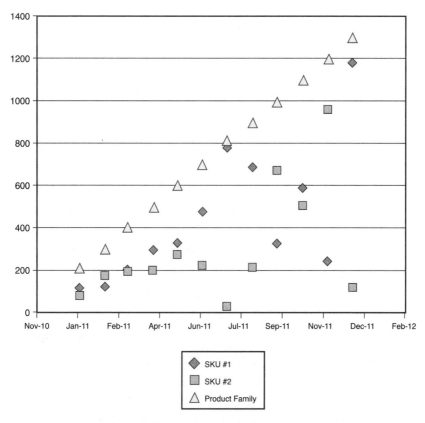

Figure 3-16 Product family level forecasting example: scatter plot

I admit that I've stacked the deck in this example to make a point: Sometimes the case is that demand at the SKU level is so random that no pattern can be detected, but if SKUs can logically be grouped into product families, patterns might emerge. What I've seen work well in practice is to perform statistical analysis at the product family level, and then apply an average percentage to the SKUs that make up the product family. In other words, if SKU #1 averages 13% of the volume for the product family as a whole, then plan production for SKU #1 at 13% of the forecasted volume for the product family. So, for the forecaster who is faced with 10,000 SKUs to forecast, perhaps there are only 1,000 product families—still a daunting challenge, but considerably more manageable.

Regression Analysis

Time series analysis is the term used to describe a set of statistical tools that are useful for identifying patterns of demand that repeat periodically—in other words, patterns that are driven by *time*. The other most widely used tool for demand forecasting is *regression analysis*. This statistical tool is useful when the analyst has reason to believe that some measurable factor other than time is affecting demand. Regression analysis begins with the identification of two categories of variables: dependent variables and independent variables. In the context of demand forecasting, the dependent variable will always be *demand*. The independent variable(s) are those factors that the analyst has reason to believe might influence demand. Consider the case of a demand forecaster at an automobile company. Identifying measurable factors that can influence demand for new automobiles is easy. Interest rates, for example, probably affect demand. As interest rates go up, demand probably goes down. Unemployment rates probably affect demand. As unemployment rates go down, demand for new automobiles probably goes up. Fuel prices might affect demand, but they might affect demand for different vehicles differently. As fuel prices rise, demand for SUVs probably goes down, while demand for hybrids probably goes up. Thus, these external, economic factors appear to be good candidates to be independent variables that might affect demand. Regression analysis that examines these types of external variables is most appropriate for forecasts at higher levels in the forecasting hierarchy (that is, product category or brand, rather than SKU-level forecasting), as well as forecasts with a relatively longer time horizon.

In addition to these external variables, which the firm has little to no ability to control, internal measurable factors, which the firm can control, also affect demand. Examples of these internal factors are promotional expenditures, pricing changes, number of salespeople, number of distribution outlets, and so forth. Any of these measurable factors are again good candidates to be considered as independent variables that might affect the dependent variable—*demand*. The term that is often used, especially in industries that are very promotional-intensive, is *lift*. Regression analysis can be very useful for documenting the lift that occurs when different types of demand-enhancing activities are executed. Understanding lift is useful for both strategic decision-making ("Is the lift from network advertising greater than the lift from cable advertising?"), and for operational forecasting ("What will the lift be from the promotion that is scheduled to run in 3 weeks?").

Regression analysis comes in many flavors, and the easiest to explain is simple, linear regression. "Simple" implies that only one independent variable at a time is being analyzed, as contrasted with "multiple" regression, in which the analyst is simultaneously considering more than one independent variable. "Linear" regression implies that the analyst is assuming a linear, rather than a curvilinear, relationship between the dependent and independent variables. The easiest way to explain how regression analysis works is to show an example of simple linear regression. Table 3-3 contains 36 months of monthly demand data for a particular product, along with monthly advertising expenditures for that product.

Table 3-3 Example of Regression Analysis

	Ad Dollars (Thousands)	Demand (Thousands)
Jan-09	200	4858
Feb-09	202	5475
Mar-09	204	5215
Apr-09	206	4884
May-09	208	5426
Jun-09	210	5342
Jul-09	212	5265
Aug-09	214	5382
Sep-09	217	5456
Oct-09	219	5367

Table 3-3 continued

	Ad Dollars (Thousands)	Demand (Thousands)
Nov-09	221	5727
Dec-09	223	5092
Jan-10	225	5356
Feb-10	228	5881
Mar-10	230	5346
Apr-10	232	5530
May-10	235	6209
Jun-10	237	5739
Jul-10	239	5870
Aug-10	242	5942
Sep-10	244	5884
Oct-10	246	6534
Nov-10	249	6165
Dec-10	251	6730
Jan-11	254	6197
Feb-11	256	6681
Mar-11	259	6287
Apr-11	262	6121
May-11	264	6278
Jun-11	267	6511
Jul-11	270	6255
Aug-11	272	6744
Sep-11	275	6664
Oct-11	278	7229
Nov-11	281	6538
Dec-11	283	7422

Figure 3-17 shows a scatter plot of these "matched pairs" of data, with each month constituting a matched pair. The vertical axis (the dependent variable, or the Y axis) is the level of demand that occurred in that particular month, in thousands, and the horizontal axis (the independent variable, or the X axis) is the level of advertising expenditure that occurred in that same month. Regression analysis tries to answer two questions from these data:

- Is there a relationship between advertising expenditure and demand?

- Can that relationship be quantified?

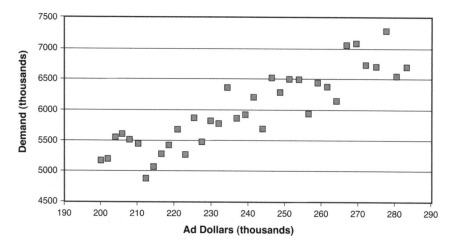

Figure 3-17 How regression analysis works: scatter plot

These questions are answered in Figure 3-18, which contains the output of a regression analysis on these matched-pair data. What simple linear regression does it to first draw a line that constitutes the best "fit" to the data. In this case, "best fit" means that the line created by the analysis is the line in which the total variance between all the data points and the line is minimized. Figure 3-18 shows that regression line. The statistics found at the bottom of Figure 3-18 provide the answers to the two questions posed earlier: "Is there a relationship between advertising expenditure and demand?" The statistics tell us that the answer is "yes." In the row labeled "Ad Dollars (thousands)," notice a number in the column "p-level" that reads 0.0001. This can be interpreted that a near-zero probability exists that advertising expenditures and demand are *not* correlated. So (thankfully), the analyst in this case can proceed with a high level of confidence that at least over the last 36 months, advertising expenditures did indeed have a relationship with demand.

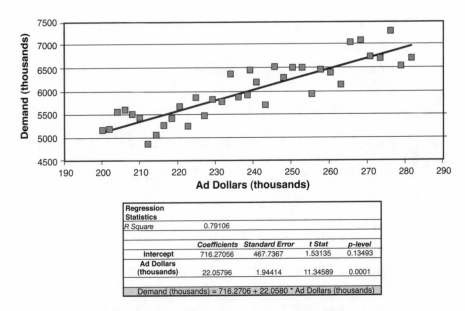

Regression Statistics				
R Square	0.79106			
	Coefficients	Standard Error	t Stat	p-level
Intercept	716.27056	467.7367	1.53135	0.13493
Ad Dollars (thousands)	22.05796	1.94414	11.34589	0.0001

Demand (thousands) = 716.2706 + 22.0580 * Ad Dollars (thousands)

Figure 3-18 How regression analysis works: draw regression line

The second question is, "Can that relationship be quantified?" Figure 3-18 again provides the answer. The statistics at the bottom of the figure contain a "coefficient" for the variable "ad dollars," which in this case is 22.05796. Because this example is simple, linear regression, you can interpret this number as the slope of the regression line that has been drawn through the matched-pairs data. In practical terms, this number means that over the past three years, every additional thousand dollars spent on advertising has resulted in an additional 22,057 units of demand for this product.

The equation at the bottom of the figure represents the overall "answer" from this exercise in regression analysis. The demand planner would like to know what demand will be for this product in, let's say, 3 months. All the demand planner needs to do, then, is to call his or her friendly advertising manager and ask, "What will advertising expenditures be in 3 months?" Suppose the advertising manager answers, "Our plan is to spend $247,000 in advertising 3 months from now." The demand planner then simply plugs that number in the equation found at the bottom of Figure 3-18:

Demand (in thousands) = $716.2706 + (22.0580 \times 247) =$ 6,164.597

Because this is expressed in thousands, then the forecast 3 months from now should be 6,164,597 units. *Voilà*, a forecast!

One must consider important caveats concerning the examples of regression analysis described earlier. First, this is once again an exercise in "looking in the rear-view mirror." The analysis being conducted here is based on the relationships between independent variables and demand *that occurred in the past*. Remember the example of a demand planner in the automobile industry who could use regression analysis to predict the effect that interest rates would have on demand for new automobiles. Regression can only tell you what that relationship has been historically. Human judgment must be applied to answer the question, "Do we think that the relationship between interest rates and demand for new automobiles will be the same next year as it was last year?" The analyst might decide that no reason exists to believe that this historical relationship between interest rates and demand for new automobiles will change in the future. But considering this question is necessary.

Another caveat involves properties that must be present in the data for regression analysis to be valid. Issues such as normality of the data, lack of autocorrelation, and heterosketasticity need to be addressed before the results of regression analysis can be considered valid. I again refer the reader to the opening paragraph of this chapter, which contains my disclaimer about going into detail about statistics. The reader can find other sources that are much better references for determining whether the data involved in these analyses conform to the statistical requirements for regression.

A final caveat involves the distinction between *correlation* and *causality*. Correlation between variables can be determined statistically using tools such as regression analysis. However, the statistics cannot answer the question of causality. In other words, referring to a previous example, do high interest rates *cause* a drop in demand for new automobiles, or does a drop in demand for new automobiles *cause* an increase in interest rates? In this example, common sense would suggest that it is the former, rather than the latter. But in other situations a third factor might in fact be "causing" both the level of demand, and the independent variable being considered, to vary simultaneously. Once again, human judgment must be applied to make sure that the statistical analyses are reasonable, valid, and actionable.

Summary

Note that statistical forecasting is not always useful. Chapter 2 offered the example of Boeing and its need to forecast demand for commercial aircraft. The point was made then that when a company has only 300 customers worldwide, and each product costs hundreds of millions of dollars to purchase, then statistical demand forecasting might not hold much value. Simply asking those customers what their demand is likely to be in the years to come might be far more useful.

But aside from these relatively rare examples, baseline statistical forecasts are an excellent way to begin the forecasting task. Looking in the rear-view mirror is often a good place to start. Understanding historical demand patterns can be extremely helpful, whether those patterns repeat with time or whether fluctuations in historical demand can be understood by discovering the relationship between demand and other factors. As discussed in Chapter 7, "World-Class Demand Forecasting," best practice involves a stepwise process, where the first step is exploring historical demand, identifying patterns, and projecting those patterns into the future. The important point to make here, though, is that this process of statistical forecasting *is just a first step.* Too often, companies fail to capitalize on the judgment of informed people to answer the question, "Are these patterns likely to continue into the future?" A risk is associated with over-reliance on statistical forecasting. When no step is in place for adding insights from sales or marketing (in a manufacturing environment), or merchandising (in a retailing environment), then the forecaster has no way to judge whether the future will look any different from the past. Another way to think of it is that if you only look in the rear-view mirror, you might get hit by a truck! On that note, the discussion now turns to the forward-looking process of qualitative forecasting.

4

Qualitative Forecasting Techniques

Chapter 3 directed your attention to looking in the rear-view mirror—to techniques for examining historical demand data, finding patterns in that historical demand, and then projecting those patterns into the future. You gain much insight from this exercise, but as concluded at the end of Chapter 3, if you only look in the rear-view mirror, you might get hit by a truck! This chapter discusses techniques and approaches for pointing your eyes out through the front windshield—at the future. In other words, it discusses the topic of qualitative forecasting.

First, the chapter defines qualitative forecasting and discusses when employing these qualitative techniques makes sense. Then, it discusses the most typical sources of qualitative forecasts, and articulates some of the advantages and problems associated with doing qualitative forecasting. Following this overview, a more detailed discussion ensues of the most common qualitative forecasting techniques, focusing your attention on the variety of expert evaluation techniques that are most commonly used. Considerable attention is devoted to the subject of salesforce forecasting, because it is the most widely used of all qualitative techniques.

What Is Qualitative Forecasting?

Simply defined, *qualitative forecasting* (also called subjective or judgmental forecasting) is the process of capturing the opinions, knowledge, and intuition of experienced people, and turning those opinions, knowledge, and intuition into formal forecasts. It is the explicit process of turning toward the future, incorporating all available information, and using that information to estimate what future demand will be. In some forecasting situations, such opinions and judgments are the best, and indeed only, source of information available about future demand. Chapter 2, "Demand Forecasting as a Management Process," provides the example of Boeing and the task of creating a demand forecast for commercial aircraft. In this case, the judgments of experienced people will be much more useful than any statistical evaluation of historical demand. However, in most situations, the best approach to forecasting demand is to augment the insights that come from statistical analysis of historical demand with the insights that come from these qualitative judgments. In other words, best practice requires the forecaster to look in both the rearview mirror (statistical forecasting) and through the windshield (qualitative forecasting) to get the best all-around view of future demand.

Demand forecasters need to employ qualitative forecasting techniques when they have reason to believe that the future will not necessarily look like the past. Typically, this takes place in three different situations. First, qualitative judgment needs to be employed when new products, for which no historical demand data exists, need to be forecasted. Historical demand data often exists for similar products that can be used to guide the forecaster in a new product situation, but the forecaster still must apply qualitative judgment to determine which "old" products can best serve as a guide, and to what extent the experience from the "old" product will, in fact, be replicated.

The second situation where qualitative forecasting is called for is a situation where new conditions are expected to change previous patterns. Take a look at a hypothetical example to illustrate this point. Hershey Foods served as an example in previous chapters, so let's use Hershey Foods again. Imagine that a forecaster is faced with the task of predicting demand for Reese's Peanut Butter Cups (a

Hershey product) during the month of October, which is a big month for candy demand (think Halloween). Imagine further that for the past several years, Hershey Foods has convinced Wal-Mart (a very big seller of Halloween candy) to put up an end-of-aisle display at the end of the candy aisle in every Wal-Mart store in the U.S. To continue with this hypothetical example, imagine that for the upcoming Halloween period, Wal-Mart has decided, for one reason or another, to put up an end-of-aisle display for one of Hershey Foods' competitors, such as Nestle, rather than Hershey's display. Given the volume of product that is sold through Wal-Mart, this new condition is likely to have a large effect on overall demand for Reese's Peanut Butter Cups. However, if the forecaster only uses statistical forecasting (that is, only looks in the rear-view mirror), he or she will presume that the demand experienced in previous years at Wal-Mart will be repeated this year. The result will be an overforecast of Reese's Peanut Butter Cups, leading to excess inventory, possible obsolescence (because this product is perishable), and heavy markdowns. The answer? Qualitative judgment must be used to supplement the insight gained from statistical forecasts. This insight could be obtained from the Hershey sales team that calls on Wal-Mart. Unless a process exists for obtaining that sales insight, and an incentive for that sales team to provide that insight, then the forecast will be wrong.

The third situation that calls for use of qualitative forecasting is one where the nature of the product is such that historical demand is really not relevant. The Boeing case from earlier is an example of this phenomenon. Another example is one used in Chapter 2—a project-based business. In this situation, bids are submitted for what are sometimes highly complex solutions to large-scale customer requirements, and the seller either wins the project, or not. Although you can use some analytical tools to track the percentage of bids that are acquired in different types of circumstances, little benefit is to be gained from the use of time-series or regression analyses in these situations. Again, the forecaster has little to gain from looking in the rear-view mirror (statistical forecasting), and much to be gained from seeking the insights of those individuals who are closest to the customer (qualitative forecasting).

Who Does Qualitative Forecasting?

Although qualitative demand forecasts can come from a variety of sources, the most common sources of qualitative insight that are internal to the company are

- Senior executives
- Marketing or product management
- Sales

Senior executives often have valuable insights about long-term trends for brands or product categories, because they typically focus on "big-picture" industry trends, market shifts, and competitive dynamics. Marketing, or product management people, have valuable insights about shorter-term promotional activities, the timing of new product introductions, and other changes to the product portfolio. Sales, which is the most frequent provider of qualitative forecasts, has valuable insights about future demand from individual customers. In the previous examples, the forecasters at Boeing would need help from the sales teams to predict demand from their customers; the forecasters at Hershey would need insight from the Wal-Mart team to understand the shift in Halloween demand; and the forecasters at the project-based business would need the judgment of their sales teams to determine the likelihood of "winning" the contracts out for bid, as well as the likely mix of products and services that those contracts would bring.

Two of the three categories of individuals who typically provide qualitative forecasts are individuals from marketing and sales. Marketing, of course, includes those people whose tasks include developing and introducing new products, building and maintaining the firm's brands, conducting market research to stay in touch with customer needs and wants, and developing and executing various promotional activities, including advertising. Sales, of course, is responsible for building and maintaining relationships with specific customers, and driving revenue from those customers. In both cases—sales and marketing—these individuals are responsible for generating and maintaining *demand*. Take a look back at Chapter 2. The focus of this book is *demand forecasting*, which is the process of estimating future *demand*. I reiterate this point here to emphasize how critical it is that sales and marketing be active participants in the demand forecasting process. No one has better insights about demand than sales and marketing, yet these groups often are either non-participants in the forecasting process, or they actively undermine the forecasting process by following an agenda that runs counter to forecasting excellence. The future sections, "Personal Agendas" and "Salesforce Composite," revisit these issues during the discussion of the politics of forecasting.

Advantages of Qualitative Forecasting Techniques

The principal, and significant, advantage of qualitative forecasting lies in its potential for predicting changes that can occur in demand patterns. Time series quantitative techniques cannot predict changes in demand patterns. Regression cannot predict changes in the relationships between demand and the independent variables that affect demand. Predicting the occurrence and nature of these changes can be accomplished by qualitative analyses based on the knowledge and experience of people both internal and external to the company. This information is valuable by itself or as additional information to be utilized to adjust the quantitative forecasts.

A second advantage of qualitative forecasting techniques is that they make use of the extremely rich data sources represented by the intuition and judgment of experienced executives, sales employees, marketing people, channel members, and other experts. Everything from customer insights, to assessment of competitive intelligence, to analysis of changing consumer trends, and even to the effect that changing weather patterns might have on demand, can be a source of insight that can lead to qualitative forecasts. That's the good news— qualitative forecasts can make use of rich data sources. The bad news, discussed next, is that forecasters can sometimes be overwhelmed with the quantity, complexity, and even conflicting messages that come from these rich data sources.

Problems with Qualitative Forecasting Techniques

A number of problems exist in the use of qualitative forecasting techniques. Some of these problems are unavoidably tied to the nature of judgmental forecasting. Other problems are due to the fact that qualitative forecasting involves collecting information from people, and people often have their own agendas that result in inaccurate or biased forecasts.

Large Amounts of Complex Information

One problem with qualitative forecasting, as mentioned earlier, is that it can include large amounts of complex information, which might very well cause the forecaster more confusion than clarity. Chapter 5, "Incorporating Market Intelligence into the Forecast," talks about the value of market intelligence as an input to the forecasting process, and this market intelligence is really a type of qualitative data that forecasters can use. Market intelligence includes external sources, such as industry reports, economic forecasts, and other macro environmental trend data. It can also include information, which might be data-oriented or insight-oriented, that comes directly from customers. It also includes internal sources, such as information from product or brand management, new product development, marketing, or sales. Any, or all of these sources might be contributing information to the forecasting process, and getting overwhelmed with the volume is easy to do for the forecaster. A major source of qualitative forecast bias is the forecasters' limited ability to process all this complex information, as well as a limited ability, or even willingness, to acquire information. Integrating numerous, complex bits of information is difficult for people. People also have a tendency to make use of information that is already available to them, or to which they have been most recently exposed. Consequently, qualitative forecasts are frequently generated without considering all relevant information, or perhaps using only that information that is readily available or has been most recently learned. Once again, it's a good news/bad news story. The good news is that the forecaster can incorporate rich sources of data into the forecast. The bad news is that a lot of information might be available, and much of it might be conflicting or confusing.

Information Limitations

Another problem inherent in qualitative forecasting is that the forecaster might be limited by the availability, timeliness, or format of the information. For example, the most common source of qualitative forecasting input comes from the organization's salesforce. However, forecasters often struggle with getting salespeople to submit their forecasts in a timely manner, for reasons discussed later on this chapter. Similarly, some external sources of qualitative input, such as industry reports or macroeconomic data, might not be as up-to-date as the forecaster would like. Finally, information might not come in the form that would be most useful for the forecaster. For example, customers might submit forecasts using their own part numbers, rather than the part numbers that correspond to the SKU-level forecasts being done, which require extra time and energy spent in translation.

Cost Issues

Another inherent problem with qualitative forecasting involves cost, both financial and time. Previously in this chapter, the point was made that the primary internal sources for qualitative forecasts are senior executives, product or brand managers, and salespeople. One thing these three groups have in common is that they are all relatively highly paid, and they all have lots of things to do in addition to their participation in the forecasting process. Forecasting can be a time-consuming activity for these individuals, especially if the process is not organized efficiently, and the time they spend on it could be spent on other value-added activities. Later, this chapter offers ways to make forecasting as efficient as possible, so these highly paid individuals can use their limited time for maximum benefit.

Failure to Recognize Patterns

A further problem with qualitative forecasting stems from the fact that when individuals are asked to look for patterns in historical demand data, they might fail to see patterns that do exist, or they might see patterns that don't exist. This is inherently problematic when a company's forecasting processes do not take adequate advantage of the insight that can be gained from statistical, or quantitative forecasting techniques, such as those discussed in Chapter 3, "Quantitative Forecasting Techniques." Some companies fail to perform these statistical analyses, either because of lack of funding for the available tools, or lack of expertise on the part of forecasters. Yet even when these statistical analyses are not performed, forecasters are asked to find patterns, and people are not as good at identifying these histori-cal patterns as the statistical tools are.

Personal Agendas

The final, and perhaps most common problem with qualitative forecasting lies in the "game playing" that often takes place when individuals with different agendas are asked to participate in the pro-cess. Ideally, when someone contributes qualitative insight to a fore-casting process, his or her goal should be to contribute to the most accurate, least biased demand forecast possible. However, individuals often have a personal or organizational goal in mind that differs from the goal of an accurate, unbiased forecast. These other agendas lead to biased forecasts. Here are some examples of the "games" that are often played by those who contribute qualitative forecasts:

- **The "My forecast will affect my quota" game.** Later this chapter covers in great detail the role that salespeople play in the forecasting process. One of the most common biases that is introduced into qualitative forecasts arises from the percep-tion on the part of the salesforce that their forecast will affect their quota, or sales target. The thought process goes as follows: "If my forecast affects my quota, then I'll bias my forecast in a downward direction. My quota will then be nice and low, and when I exceed my quota, then I'll look like a hero and make a big bonus."

- **The "I'll forecast high so I'll get what I need" game.** Salespeople's forecasts will have an upward bias if they believe, either correctly or incorrectly, that the goods or services that they sell are capacity constrained. The thought process will then be something like, "I'm only going to get 80% of what I forecast, so if I forecast 125% of what I think my customers will want, then I'll get all I need." With this "game," the qualitative demand forecasts will inevitably be biased on the high side.

- **The "My new product won't get through the New Product Development (NPD) process if my forecast fails to be high enough" game.** One important input to qualitative forecasts is the judgment of future demand for new products. Product or brand managers often feel pressure to get new products into the marketplace, and they know that NPD processes usually have a business analysis "hurdle" that must be crossed, where business leaders examine the forecasted demand for the new product. Product managers know their new product won't get the "green light" if the demand forecast is too low. This "optimism agenda" leads to upward bias in many new product introductions.

- **The "I need to make sure they make enough, so I'll forecast a lot" game.** One source of qualitative input is downstream members of the firm's supply chain—in other words, customer forecasts. Whether it is manufacturers that are downstream from raw material or component suppliers, or wholesalers or retailers that are downstream from manufacturers, upward bias might be introduced into their forecasts through an either conscious or unconscious effort to make sure the upstream supplier has enough product to meet their needs. At one company we worked with, this practice was described as "uncommitted commitments." The customer would "commit" to buying a certain amount of product as a way to encourage the manufacturer to "make enough," but these commitments were not firm orders. Those firm orders often didn't materialize, and these "uncommitted commitments" resulted in forecasts that were biased high.

Summary: Qualitative Technique Advantages and Problems

Table 4-1 provides a summary of the advantages and problems associated with qualitative forecasting. Despite this rather long discussion of the problems associated with qualitative techniques, keep in mind that qualitative techniques are a valuable resource for any forecaster. The value of experience and the ability to analyze complex situations as input to an overall demand forecasting process should never be discounted. Indeed, every forecast involves some degree of qualitative input. The discussion of the problems associated with qualitative techniques was presented here solely for the purpose of helping you make better qualitative forecasts by avoiding some of the common "traps" associated with these techniques. With these traps in mind, the following section moves on to a discussion of the qualitative techniques available.

Table 4-1 Summary of Advantages and Problems with Qualitative Forecasting

Advantages	Problems
Can predict changes in historical demand patterns	Large amounts of complex information that may be conflicting or confusing
Can incorporate very rich sources of data	Limited by the availability, recency, or format of the data
	Can be expensive and time consuming
	Individuals might fail to see patterns that exist, or see patterns that don't exist.
	Qualitative forecasts are subject to "game playing."

Qualitative Techniques and Tools

This section covers several qualitative forecasting techniques that use the judgment, knowledge, and intuition of experienced people to produce and enhance demand forecasts. The techniques discussed solicit expert evaluations via the jury of executive opinion, the Delphi method, and salesforce composites. These expert evaluations use the experience of people, such as executives, salespeople, or marketing people, who are familiar with a product line or a group of products, to generate demand forecasts. The techniques in this section generally involve combining inputs from multiple sources; that is, groups of executives, salespeople, or marketing people. The advantage of soliciting contributions from more than one person, of course, is that it can offset biases introduced into a forecast when one person produces the forecast. This section focuses on the techniques for collecting and organizing qualitative insights from people internal to the firm. Chapter 5 broadens the discussion to include the collection and analysis of external information, which is referred to as *market intelligence*.

Jury of Executive Opinion

When executives from various corporate functions involved in demand generation—primarily sales, marketing, or product management in a manufacturing environment, along with merchandising in a retail environment—meet to generate or discuss forecasts, the meeting is termed a *jury of executive opinion*. It is a relatively simple forecasting technique to implement, and it is quite valuable when changes in existing demand patterns are anticipated or when no historical demand data is available for quantitative forecasting analyses. It also has the advantage of making use of the rich data represented by the intuition and judgment of experienced executives.

We have found that one of the most widespread uses of a jury of executive opinion is in a consensus forecasting process. Chapter 1, "Demand/Supply Integration," discussed the role of demand forecasting in the firm's overall Demand/Supply Integration (DSI) process. The demand review is an integral element of this process, and when the demand review is executed correctly, it becomes a consensus forecasting process. Chapter 8, "Bringing It Back to Demand/Supply Integration: Managing the Demand Review," covers the detailed elements of a demand review in greater detail, but as a preview, it should consist of representation from all the demand-generating functions in the firm, who come together to arrive at consensus on the demand forecast. In essence, this demand review becomes a jury of executive opinion. A typical process consists of quantitative demand forecasts being generated in advance of the demand review, and the consensus forecasting group meets to decide whether and how much to adjust the quantitative forecasts. Frequently, these consensus-forecasting groups are also responsible for generating qualitative forecasts for new products. The effective use of the jury of executive opinion technique depends on the degree to which the organization is able to overcome the sources of bias inherent in individual and, particularly, group decision making. To the extent that these pressures constrain the decision-making process, biased forecasts will result.

A frequent source of bias in a jury of executive opinion is political pressures within the company, usually in the form of influence exerted by the member of the jury whose function is the most powerful within the culture of the company. Because of this influence, the contributions from other members of the jury carry relatively less weight in the final forecasts. An excellent example of this influence was seen when our team audited a privately held, family-owned company that manufactured products sold through home improvement retailers. The senior sales and marketing executive, who happened to be one of the family members who owned the company, took personal responsibility for the relationship with the company's largest customer, a large home improvement "big box" retailer. Nearly 50% of this company's revenue came from this one mega-retailer. In the monthly demand review meeting, designed to arrive at a consensus forecast, this executive completely dominated, and encouraged very aggressive forecasts of demand coming from this large customer. He based his forecasts on his own inherently optimistic personality, and his "gut feel" about how much the customer would order. The result was a forecast that was consistently biased high, across all the company's product lines, and by the time we came in to perform our audit, the company was drowning in inventory. The advice that we offered this company was to focus on measurement. We encouraged their demand planning team to collect data on the accuracy of the forecasts *before* this optimistic executive adjusted them, and compare them to the accuracy of the forecasts *after* he made his adjustments. The team then showed this executive a chart that clearly displayed the correspondence between rising inventory levels and his biased forecasts. Because he partially owned the company, he got the message!

Another risk that is common with jury of executive opinion forecasting is the creation of *plan-driven forecasting* (refer to Chapter 1). Sales and marketing executives are often overwhelmingly conscious of the business plans, or financial targets, that they must reach. Our work with companies has shown that frequently, these demand-side executives attempt to influence the consensus forecasting process to ensure that the resultant forecast is in alignment with the financial goal, even when the rigorous forecasting process uncovers no evidence that sufficient demand exists in the marketplace to achieve this financial goal. As discussed in Chapter 1, plan-driven forecasting is perhaps the most insidious type of aberration to the ideal of DSI, and without the discipline that comes from strong metrics, jury of executive opinion techniques can lead to this negative result.

An important caveat to the use of the jury of executive opinion is that the technique is not appropriate for short-term forecasts at very granular levels of the forecasting hierarchy, such as SKU-level forecasts. A jury of executive opinion, by its very nature, requires valuable executive time; therefore, the most efficient use of this technique is for strategic-level forecasts for groups of products; that is, product lines or product families. Using a jury of executive opinion for low-level, short-term forecasts encourages bias because of the repetitive nature of these forecasts and is a waste of costly executive time.

Companies using a jury of executive opinion in their forecasting process should also be aware that this technique tends to disperse responsibility for forecasting accuracy. We have found that unless companies using a jury of executive opinion are relatively sophisticated in managing their forecasting process, members of the jury are neither evaluated, nor rewarded, for forecasting accuracy. When no one has accountability for forecast accuracy, inaccurate forecasts inevitably result. Companies that use a jury of executive opinion successfully do so, in part, because they both evaluate and reward members of their consensus forecasting group for forecasting accuracy.

Another procedure that can be used to assign responsibility for accurate forecasts when using a jury of executive opinion is to require written justification for qualitative adjustments to quantitative forecasts. When this documentation is required, it not only has the effect of assigning responsibility for accurate forecasting, but it also makes performing post-hoc analyses easier. In other words, if forecasts prove to be inaccurate, the documentation makes determining the reasons for the inaccuracies easier. A best practice in conducting demand review meetings is to have detailed notes taken, thus documenting the logic behind any adjustment to the quantitative forecasts, or forecasts for new products.

Delphi Method

A *Delphi method* is a procedure used to collect the opinions of knowledgeable experts, either internal or external to a company, that attempts to minimize interpersonal biases that sometimes accompany a jury of executive opinion. A Delphi procedure utilizes the following steps:

1. Each member of the panel of experts who is chosen to participate is presented with a strategic question involving a forecast of future demand. For example, a group of experts at Boeing might convene a Delphi panel to answer a question like, "What

will demand be for 747-class airplanes in 15 years?" This panel of experts might include senior executives at Boeing, senior sales executives, and even external experts who study the airline industry.

2. Each member of the panel independently ponders this question, and then writes an answer, along with a detailed discussion of all the reasoning behind his or her answer.

3. The answers provided by the panel members, along with their rationale, are given to a scribe, who transfers these responses into a single document. This document is then returned to the members of the panel, but without the identification of which expert came up with each forecast.

4. After reading the summary of replies, each member of the panel either maintains his or her forecast or reevaluates the initial forecast and submits the new forecast (and the reasoning behind changing his or her forecast) in writing.

5. The answers are summarized and returned to panel members as many times as necessary to narrow the range of forecast. Eventually, the panel tends to converge on a forecast.

An appropriate use of the Delphi method is for the prediction of mid- to long-term company or industry demand levels. When this technique is used within a company, one can think of it as a kind of "virtual" jury of executive opinion, because the executives do not meet face to face. The purpose of this distance is to allow each member to use his or her reasoning to develop a forecast without the influence of strong personalities or the fact that the "boss" has a pet forecast.

The Delphi method also reduces the effects of "groupthink" on the decision-making process. Because the participants do not meet face to face, the bias that occurs because of a desire on the part of group members to support each other's positions or the influence of a strong leader within the group is minimized. Removing this source of bias enables conflicting ideas to survive long enough to be examined, thus allowing a range of scenarios to emerge from the process and an outcome that is more legitimate, particularly when long-term demand forecasts are being made.

Problems with this method of qualitative forecasting focus on its tendency to be unreliable; that is, the outcomes can be highly dependent on the composition and expertise of panel members. To some extent this source of bias is the result of group members not being willing or able to seek out information other than what is readily available or recently acquired. Supplying panel members with relevant information (for example, economic or industry indicators) can reduce this source of bias. In addition to this bias, the Delphi method is very time consuming and thus expensive. Such a technique is most appropriate for long-term, strategic-level forecasts rather than short-term, operational ones.

Salesforce Composite

The salesforce composite is a qualitative forecasting method that uses the knowledge and experience of a company's salespeople and its sales management to augment or produce demand forecasts. The grass roots approach to a salesforce composite accumulates demand forecasts for the regions, products, or customers of individual salespeople. The sales management approach seeks forecasts from sales executives and is essentially a jury of executive opinion, albeit consisting of a narrower range of executives (that is, only sales executives). A recent article provides a summary of a survey of forecasting practices by salespeople.[1] Key findings from this reported survey include

- Almost 82% of salespeople surveyed participate in forecasting.
- At the same time, only 14% of salespeople receive training in forecasting.

[1]   McCarthy, Teresa M., Mark A. Moon, and John T. Mentzer (2011), "Motivating the Industrial Sales Force in the Sales Forecasting Process," *Industrial Marketing Management*, 40 (1), 128–138.

- Almost half (more than 47%) of salespeople report that they have either no, little, or some knowledge of what happens to their forecasts after they are submitted.

- Only 16% of salespeople have access to forecasting software to assist them in their forecasting tasks.

- Less than half of the salespeople believe that the quality of their forecasts affects their performance evaluations.

The picture painted from this survey is that although an overwhelming majority of salespeople are responsible for forecasting, a considerable gap exists between the expectations that companies have for them and the resources that companies provide them to excel at this critical task.

Despite this gap, important advantages exist for the salesforce composite forecasting technique. It has the potential for incorporating the expertise of people who are closest to the customer. In addition, the technique places forecasting responsibility on those who have both the ability to directly affect product sales and the potential to experience the impact (in the form of their customers' displeasure, for example) of forecasting errors.

Two general situations call for the salespeople to participate in a company's forecasting efforts:

- **When salespeople manage ongoing streams of product flow to their customers, be they end-use customers or channel partner customers.** In these situations, salespeople are the most natural sources of information regarding changes to patterns of demand. Remember the hypothetical example of Hershey Foods and demand from Wal-Mart for Halloween candy earlier in this chapter? In this hypothetical example, Wal-Mart decided to change its historical pattern of ordering from Hershey in advance of Halloween. As discussed, without information flowing to the forecasting team about these changing patterns, overforecasting would result. This information has to come from the sales team.

- **When salespeople work with large project or proposal-based sales.** In this case, accurate forecasts require the intelligence that salespeople have concerning the likelihood of securing large orders. For example, a computer company that sells large data processing systems needs a prediction of the likelihood of winning a large contract from a major customer. If such a win is likely, then the computer company needs to adequately plan for the increased demand. The salespeople are in the best position to assess that likelihood.

Although salespeople provide critical input to many forecasting processes, companies are frequently frustrated by the quality of the input that salespeople provide.[2] However, companies can do a number of things to improve the quality of salesforce input to the forecasting process.

Make Forecasting Part of Their Jobs

The first, and perhaps most important, change that companies can make to enhance salesforce forecasting is to *make it part of their jobs*. At many companies with which we have worked, salespeople make comments like, "Why should I spend my time forecasting? I've been hired to sell, not to forecast!" However, salespeople are responsible for three main activities: to sell products and services, to build and maintain relationships with their customers, and to provide market intelligence back to their companies. One of the most important forms of market intelligence is intelligence concerning future demand—in other words, forecasts. Although most sales executives would agree that these are the critical tasks they expect salespeople to perform, in many cases, salespeople are measured and rewarded for only one of

---

[2] The following discussion on improving salesperson forecasting is drawn largely from Moon, Mark A. and John T. Mentzer (1999), "Improving Salesforce Forecasting," *Journal of Business Forecasting*, 18, (Summer), 7–12.

those tasks: selling and generating revenue. Chapter 6 emphatically makes the point that *what gets measured gets rewarded, and what gets rewarded gets done.* Thus, if salespeople are not measured and rewarded for forecasting performance, they will not perceive it as part of their jobs.

How can companies make forecasting a recognized part of a salesperson's job? A first step should be to explicitly emphasize forecasting responsibilities in a salesperson's formal job description. But beyond that, forecasting must be included in the performance evaluation process and compensation strategy for the salesforce. Companies should adopt some of the performance measurement strategies discussed in Chapter 6, and these measures should be applied to salesforce forecasts. I am by no means suggesting that forecast accuracy should be the primary measure for salesperson success or failure. However, it should be a part of a "balanced scorecard" for members of the salesforce, and forecasting performance should receive enough weight on that scorecard that it gets the attention and effort needed to do a good job.

In addition, salespeople must receive training to enhance their forecasting skills. Training is a normal part of most salespeople's jobs, yet that training seldom includes forecasting training. Topics for salesforce training should include the role of quantitative forecasting, how forecasts are used by other functions in the company, and how to work with customers to convince them that accurate forecasts are in the best interests of all parties in the supply chain. In addition to training, salespeople must receive feedback on their performance. Salespeople cannot possibly improve their performance unless they know whether their forecasts tend to be high or low, and by how much. Such feedback is a critical part of helping salespeople recognize how important forecasting is to their organization.

Minimize Game-Playing

Another important emphasis area for a company to enhance the effectiveness of salesperson forecasting is to minimize the "game-playing" described earlier in this chapter. Game-playing can result in bias in either an upward or downward direction. Upward bias most frequently occurs when salespeople perceive that supply of goods and services might be limited, and they intentionally inflate forecasts to ensure receiving adequate supply for their customers. Downward bias most frequently occurs when salespeople perceive that forecasts influence quotas, or sales goals. Companies can minimize both of these types of bias. Constantly measuring forecast accuracy so that either form of bias is identified is critical. When such measurement occurs and feedback is given, then bias can, over time, be reduced. Separating forecasts from quotas in the minds of salespeople is also critical and can be done in a variety of ways. One way is to encourage salespeople to forecast in physical units (the most useful type of forecast for downstream planning purposes) while quotas are assigned in dollars, points, or some other unit. Another way is to assign quotas quarterly or annually, but to make forecasting a normal part of a salesperson's monthly, or in some cases weekly, job assignment.

Keep It Simple

Another key strategy for enhancing the effectiveness of salespeople's forecasting efforts is to *keep it simple*. One observation that we have made after working with dozens of companies and their salesforces is that salespeople are generally not very good at forecasting. However, they can be *very* good at adjusting forecasts. The best way for a company to "keep it simple" for their salespeople is to provide them with an initial forecast, generated through the statistical approaches discussed in Chapter 3, which they can then adjust. What we have seen in world-class forecasting companies is a process whereby time series and regression models are employed to generate quantitative forecasts, which are then provided to the sales staff for them to review, often with their customers, and make adjustments based on what they know about expected changes to previous demand patterns. When salespeople are ineffective is when they are given a "blank piece of paper" and expected to generate initial forecasts on their own. Whenever possible, companies should use salespeople as adjusters, not forecasters.

A question that often arises in this discussion of salesperson forecasting is, "When should I make an adjustment?" Figure 4-1 gives a simple framework for how a salesperson should think about adjusting a statistically generated forecast. The figure illustrates a 2×2 matrix, with the horizontal axis labeled as "Certainty" and the vertical axis labeled as "Impact." Two of the four quadrants are easily interpreted. When a salesperson is highly certain that a change in demand patterns will occur, and when the impact of that change is high, then he or she should definitely adjust the forecast to account for this change. The other easily interpreted quadrant is a situation where the salesperson perceives that a change in demand patters might occur, but the certainty level is low, and the impact of the change is also low. In this case, the salesperson should leave the statistically generated forecast alone, and make no adjustment. The more challenging quadrants are those where certainty is low and impact is high, or where certainty is high and impact is low. In these situations, the salesperson must apply careful judgment, considering the overall effect of either over- or under-forecasting, and decide accordingly. If feedback is provided post-hoc, then the salesperson can learn from that feedback whether the judgment made was the correct one, which will help him or her be more skilled in making future decisions of this sort.

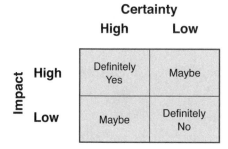

Figure 4-1 When should salespeople make adjustments to statistically generated forecasts?

Keep It Focused

A final strategy that companies can use to enhance salesperson effectiveness is to *keep it focused*. By this, we mean that for most salespeople, the "80/20" rule is a reality along two dimensions: customers and products. In other words, 20% of a salesperson's customers generate 80% of his or her business. Similarly, 20% of a salesperson's product portfolio generates 80% of his or her business. When either or both of these concentration principles are in effect, then those salespeople should be forecasting only those 20% of customers, or 20% of products, that generate the bulk of their business. If a salesperson has 100 total customers, and 100 products in his or her portfolio, then that salesperson would theoretically be responsible for 10,000 forecasts. But when a company "keeps it focused," that salesperson might only pay attention to his or her top 20 customers and top 20 products, resulting in a forecasting workload of (at most) 400 forecasts per month. Such a process has several advantages. For one thing, when salespeople perceive that the magnitude of the forecasting job is enormous, they are likely to resist and do a poor job on all forecasts. Also, salespeople are likely to have very limited information on those 80% of customers and/or products that do not generate significant revenue. When they are forced to provide forecasts in those situations where they have limited information, they are likely to turn in forecasts that are simply not very good. The bottom line is that salespeople should forecast *only* those customer/product combinations where they can really add value. In one chemical company that has participated in the audit research, the goal is for salespeople to look at, and think about, around 10% of the customer/product combinations for which they are responsible. For this company, salespeople do not see the forecasting task as onerous, and they provide excellent insights that enhance the overall accuracy and effectiveness of forecasts.

The bottom line concerning salesforce composite forecasting is that when companies use some of these strategies to enhance the effectiveness of salespeople's forecasts, those companies can increase accuracy significantly. Remember, a forecast is a best guess about what customers will demand in future time periods, and no one is closer to customers than salespeople. If companies *make it part of their jobs, minimize game playing, keep it simple, and keep it focused*, then salespeople can greatly enhance the overall forecasting process.

Summary

This chapter focused on the use of qualitative techniques that turn the opinions of experienced people into formal forecasts. The information presented included an overview of the advantages inherent in qualitative forecasting analyses, with the discussion of problems focusing on the sources of bias that influence the effectiveness of these forecasts. Qualitative forecasting techniques that were discussed as methods for tapping the knowledge and intuition of experts included jury of executive opinion, the Delphi method, and salesforce composite. In addition, this chapter presented a number of tools that are important adjuncts to the qualitative forecasting process, primarily because of their ability to enhance qualitative forecasting decisions through the reduction of the effects of the biases that can affect the accuracy of qualitative forecasts. The primary focus of this chapter has been primarily on those qualitative sources of information that are internal to the firm, such as sales, marketing, product management, and senior executives.

The bottom-line conclusion that you should draw from this discussion of qualitative forecasting is that the use of qualitative techniques *to supplement quantitative techniques*, generally improves forecasts. However, except under specialized situations such as project-based businesses, using qualitative techniques *on their own* is generally a bad idea. Finally, all qualitative techniques are more effective when they are accompanied by a rigorous process of performance measurement and regular feedback.

From here, the discussion turns to another category of information tools that can be extremely useful for making forecasts more accurate, as well as helping to manage the entire forecasting process—the use of external market intelligence as a source of forecasting knowledge.

5

Incorporating Market Intelligence
into the Forecast

As a marketing professor, I am constantly subjected to the good-natured needling from my supply chain management colleagues. One of my favorite Dilbert cartoons constitutes one of their favorite pieces of ammunition. In this strip, Dilbert makes the observation that "Marketing is liquor and guessing." Well, I made the point in Chapter 2, "Demand Forecasting as a Management Process." that forecasting is, at its core, guessing about the future. I won't agree that liquor is involved, but it is certainly the case that the "guessing" that is inherent in demand forecasting can be upgraded to "informed guessing" by the inclusion of market intelligence as a part of the forecasting process. That is the subject of this chapter—what types of market intelligence will benefit the forecasting process, and how that market intelligence can be incorporated effectively. The chapter begins by defining market intelligence, and categorizing the different types of market intelligence that can be helpful to the forecasting process. Companies can benefit from both internal sources of market intelligence—sales and marketing—and external sources of market intelligence—including customers. This chapter explores each of these sources and covers the advantages and risks of incorporating each type of intelligence. The chapter concludes with comments about how to consolidate all these different sources of information, both quantitative and qualitative, into a final forecast.

What Is Market Intelligence?

Here is a simple definition for market intelligence:

Market intelligence is a procedure for systematically gathering, interpreting, and disseminating insight about the market environment, for purposes of enhancing organizational decision-making.

Okay, fine, but why talk about market intelligence in a forecasting book? Chapter 2 defined *demand forecasting* in the following way: *A demand forecast is a firm's best estimate of what demand will be in the future, given a set of assumptions.* In essence, market intelligence is the task of collecting and analyzing information about the environment so that the demand forecaster can document the *assumptions* that underlie the forecast. This is the information that the forecaster needs to determine how the future is likely to look different from the past. The discussion on quantitative forecasting concluded that these statistical techniques were designed to identify patterns in historical demand that can then be projected into the future. But it also concluded that very often, the future will not look like the past, and we need insight into how, and why, the future will look different—and that is the role of market intelligence.

The task, then, for the demand planner, is to identify useful and credible sources of this market intelligence. Sources are both *internal* and *external* to the firm. Chapter 4, "Qualitative Forecasting Techniques," covered at some length some of the internal sources of information that forecasters attempt to exploit, such as sales, marketing, product management, and senior executives. External sources can include market research reports from third-party providers, government reports such as census data, academic research, and other types of published information. Another important source of external data is customers, and this chapter explores customer-provided data in detail.

Bottom-Up versus Top-Down Forecasts

Chapter 7, "World-Class Demand Forecasting," covers best practices in demand forecasting, as well as a concept known as the *forecasting point of view*. Just to preview that concept, two points of view exist from which you can derive a forecast. The first is called "bottom-up" forecasting. In this approach, the forecaster essentially thinks about individual customers, or individual products. What will demand from customer A be next quarter? How about customer B? How about customer C? After all customers are considered—and often, this means all large customers are considered individually, then all small customers get bundled into an "all other" pile—the forecaster simply adds them up. The total forecast is the sum of the forecasts for each customer. A forecaster can follow a similar approach for a bottom-up product forecast. What will demand for SKU 1 be? How about SKU 2? How about SKU 3? After all the SKUs are considered, then you add them up, and get a forecast. The second point of view is a "top-down" look at future demand. When considering the problem from this perspective, the forecaster first attempts to forecast *industry* demand for an entire category of products. Then, the task is to forecast the company's expected share of that industry demand. Multiply total forecasted industry demand by the firm's forecasted share, and you get a top-down forecast. The point I make in Chapter 7 is that although both perspectives have their limitations, best practice entails using *both* points of view, and then doing a detailed reconciliation when the bottom-up number is materially different from the top-down number, which it often is.

Different categories of market intelligence contribute to each of these perspectives. One way to categorize sources of market intelligence is to think of them as broadly being either "macro" or "micro." Micro-level sources of market intelligence are those that would be useful for creating a bottom-up forecast, and can be thought of as either customer based or product based. Macro-level sources are those that would be useful for creating a top-down forecast, and can be thought of as industry based. Table 5-1 provides examples of both information that is needed at both micro- and macro-levels, as well as possible sources of that market intelligence. This table is not intended to be comprehensive, nor is it necessarily even "correct" for each reader's industry. It is intended to give examples of the types of macro- and micro-level information that demand forecasters need, and possible sources of that information. Each situation will, of course, be unique.

Table 5-1 Micro versus Macro Market Intelligence

Micro Information—Bottom-up Forecast Oriented		Macro Information—Top-down Forecast Oriented	
Needed Information	*Source of Information*	*Needed Information*	*Source of Information*
Future pricing changes	Marketing or product management	Competitor activity	External market research
Promotional activity—timing and level	Marketing	Economic trends—global, regional, local	Published economic studies, staff economic team
New products—cannibalization or complement	Product management, NPD team	Industry growth (decline)	Published industry reports
Phase-in/phase-out activity	Product management, NPD team	Legislation or political influences	Published political analysis
Project business—probability, timing, and configuration	Sales	Project business—market size and competitive activity	Competitive analysis
Customer growth (decline); customers won/lost; customer inventory trends	Sales	Channel growth (decline)	Published industry reports
Projected new distribution	Sales, marketing	Entering or leaving entire markets	Industry analysis

What Do Demand Forecasters Need to Do?

It's easy to make the assertion: "Forecasters need market intelligence." The challenge is to figure out how to get it, and what demand forecasters need to actually do to acquire and use market intelligence to make their forecasts more valuable. Following are some strategies that demand forecasters need to adopt to make market intelligence a critical part of the demand forecasting process.

- Identify what information you need, and who is likely to have that information. In other words, take Table 5-1 as a starting point, but adapt it to your specific situation. Remember, both "What do I need?" and "Where will I get it?" are critically important, and potentially equally challenging, questions.

- After you identify those sources of information, create linkages with those people whose job it is to collect that information. In most organizations, individuals exist whose job is collecting and monitoring external, macro-level industry and market information. Further, in nearly all organizations, sales and marketing people are constantly acquiring insights about their customers and can provide micro-level customer and product information. Not only must these people be identified, but they must be incentivized to work with the demand forecasters to translate their macro- or micro-level insights into information that can be incorporated into demand forecasts. That was an easy sentence to write, but accomplishing its task is a hard! Refer to the discussion in Chapter 4 on obtaining qualitative forecasting information from sales and marketing.

- Establish a routine for documenting macro-level trends. The micro-level information should be contributed on a regular basis as a part of the qualitative judgment element of a demand forecasting process. However, companies often struggle with creating—and sticking with—a routine for bringing macro-level information into the forecasting process. These macro-level trends often change slowly, and covering the same ground in every demand review meeting can feel unduly repetitive. Some companies have a routine for quarterly reviews of macro trends, rather than as a part of the monthly drumbeat of the DSI process. Another direction taken by companies is to report macro trends only in terms of exceptions. In other words, the demand forecasters collect and monitor macro-level information monthly, but report only changes that are material to the forecasts at the demand review meeting.

• Document and validate the macro-level information that is received, especially from external sources. Not all macro-level information is as unbiased or accurate as it could be, and the firm would be loath to make strategic resource allocation decisions based on information that turned out to be questionable. The demand planners might very well be challenged at the demand review, so their establishing a process for validating the information received from external sources is advisable.

In summary, then, remember the definition of a demand forecast: *A demand forecast is a firm's best estimate of what demand will be in the future, given a set of assumptions.* Think of the process of collecting, interpreting, and disseminating market intelligence as the way forecasters piece together the *assumptions* that underlie every demand forecast. From here, then, the discussion turns to a special case of external market intelligence: customer-generated forecasts.

Customer-Generated Forecasts

Customer-generated demand forecasts can be an excellent source of market intelligence. Manufacturers have a variety of customers from whom they could receive forecasts. The types of customers, and the types of forecasts, include the following:

- **Project-based customers.** As discussed in Chapter 3, "Quantitative Forecasting Techniques," many companies have customers that issue RFPs (Requests for Proposals) for complex, multi-period, multi-product projects. For example, Honeywell Corporation has a division called HBS, which stands for Honeywell Building Solutions. This division partners with customers to provide a variety of systems that are needed to maintain the safety, security, and energy needs of large buildings or campuses of buildings. HBS bids for contracts to partner on large-scale building projects to provide these types of support services. A huge variety of products and services might be involved in the projects. After a project is awarded to HBS, the demand forecasters would benefit greatly from collaboration with the customer to forecast the timing and the mix of products that will be required as the project progresses, as well as the quantities and configurations of products that will be needed.

- **OEM customers.** Companies often supply raw materials or components to their customers, who then use those products as a part of their own manufacturing processes. For example, Michelin Corporation provides tires to automobile manufacturers around the world. Demand forecasters at Michelin would greatly benefit from receiving production schedules from their OEM customers, specifying the timing and the mix of tires they will require.

- **Distribution and retail customers.** Companies such as Hershey Foods sells most of its products through distributors or retail customers such as Walgreens. Demand forecasters at Hershey can benefit greatly from capturing end-user demand through point-of-sale (POS) data, which after netting out inventory on hand, can help them to forecast both sell-in (demand from the retailer) and sell-through (demand from consumers) demand. In addition, demand forecasters at companies such as Hershey Foods benefit from access to promotional calendars that can be used to predict store-level promotional activity, and the resulting spikes in demand.

Thus, regardless of the type of customer, benefits come from receiving customer-generated forecasts.

However, a number of issues must be addressed concerning customer-generated forecasts, including the following:

- Should we get forecasts from customers?
- If we do decide we want forecasts from customers, which customers should we work with?
- How should this forecasting customer collaboration take place?
- How should we incorporate the customer-generated forecasts into our process?

Should We Get Forecasts from Customers?

Receiving forecasts directly from customers has clear benefits. Who knows better what customers are likely to want to buy than the customers themselves? If customers and suppliers find themselves in strong, collaborative relationships, then collaborative forecasting can help both companies realize significant benefits in reduced inventories, increase fill rates, and lower costs. Chapter 1 first introduced these benefits, during the discussion of demand/supply integration across the supply chain. Figure 1-2 is reproduced here as Figure 5-1, but with questions raised about the risks or concerns associated with using customer-generated forecasts. Whenever benefits exist, usually risks or concerns exist as well, which are illustrated in Figure 5-1. The first question that a forecaster must address, then, is "Do the benefits that can be realized from receiving forecasts from customers exceed the costs?" Following are some of the risks, as shown in Figure 5-1:

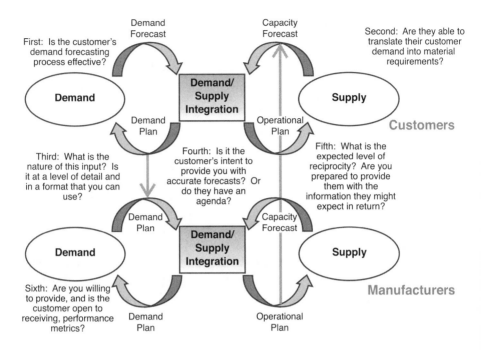

Figure 5-1 Demand/supply integration across the supply chain: the risks

1. Is the customers' demand forecasting process effective? Even if their intentions are good, the customer might not yet have read this book, and therefore, their forecasting efforts might not lead to accurate, credible forecasts! If the customer forecasts are neither accurate nor credible, then adding their input to your forecasting process might do more harm than good.

2. Can your customers effectively translate *their* customer demand into clear requirements from their suppliers? This point gets at the effectiveness of the customer's DSI processes. As we have said repeatedly, without a good DSI process in place, an accurate forecast isn't worth very much. That forecast eventually needs to be translated into good business and supply chain decisions, and if the customer's DSI processes are lacking, then the signal back to the supplier in terms of a demand forecast might be flawed.

3. Is the nature of the input that you receive from customers useful? Each company operates with different nomenclature, part numbers, and product hierarchies, and there will inevitably be some translation from the customer's nomenclature to yours. The relevant question becomes: is the customer provided forecast in a format or level of detail that makes translation of their forecast into your forecast more trouble than it's worth?

4. Is it the customers' intent to provide you with an accurate forecast of their demand? Or do they have an agenda? Chapter 4 talked about "game-playing," or the fact that individuals who contribute to demand forecasts often have "other agendas" that might lead them to deliver something other than an accurate forecast. One agenda discussed is the "I'll forecast high so I'll be sure to get what I need" game. Customers can certainly have an incentive to adopt this agenda in regard to their demand forecast. As is the case with salespeople, if customers perceive an availability problem might occur, then they might very well forecast high, hoping that the supplier will "make more" and thus avoid any shortages. One of the tricky elements of a customer-provided forecast is that this forecast is seldom perceived by the customer as a *commitment*. Rather, just like any forecast, it is a "best guess" about what will happen in the future. Because it is not a commitment, if the manufacturer builds what the customer has forecasted, and the customers then don't then buy at the level they forecasted to buy, the manufacturer is the one who is left with inventory, not the customer. Thus, often little downside risk exists to the customer from providing an inflated forecast.

5. If you are asking customers to expend effort on your behalf (that is, providing you with a demand forecast), do they understand the benefit to them? Will they ask for something in return, such as discounted pricing or guaranteed supply? This is a subject upon which the demand forecaster and the sales organization

can really collaborate. The demand forecasters can help the salespeople to position this request as a win-win proposition. The clear benefit to the customers comes from the enhanced product availability that comes from better demand forecasting.

6. As Chapter 6 discusses, forecasting performance is improved when feedback on accuracy and bias is provided to those completing the forecasts. Customer forecasts will be more useful if the customer is open to receiving feedback, in terms of accuracy and bias metrics. The question here is twofold. Is your firm willing to expend the effort required to provide this feedback to the customer, and is the customer receptive to receiving this feedback? Depending on the nature of the relationship, opportunity might exist to provide the customer with incentives for providing high-quality demand forecasts.

The bottom line, then, is that forecasters must perform a cost-benefit analysis on the question of "Should we ask our customers for demand forecasts?" They must assess the risks such as those in the preceding list, and include those risks in their cost-benefit calculation.

If We Do Decide We Want Forecasts from Customers, Which Customers Should We Work With?

Although important benefits might be available from acquiring customer-generated forecasts, an inevitable cost exists. Adding customer input to the forecasting process also adds complexity. The likelihood is that the cost in additional complexity will at some point outweigh the potential benefits. The forecaster, then, must decide on which customers to pursue for customer-generated forecasts, and which not to pursue. How do you decide?

One way is to perform a simple Pareto analysis. For most organizations, a concentration principle is in place; where 20% of customers generate 80% of the business. In many situations, that concentration level is even greater. Many companies would find it quite straightforward to name their top 20, or 50, or 100 customers in terms of business volume, and a simple way to decide on which customers should be targets of opportunity for customer-generated forecasts is to simply pursue those "A" level customers. It is often the case, however, that critically important customers can be determined by factors other than current sales volumes. Some customers have strategic importance due to their future sales potential, high margin, or some other criteria. However determined, a "first cut" at which customers should be pursued for customer-generated forecasts are those customers who are determined to be strategically important customers.

Unfortunately, it might not be that simple. Just because a customer is an "A" level customer does not necessarily mean that they are a good customer to provide customer-generated forecasts. Additional analysis must be performed. For example, a customer might be high volume, but might be terrible forecasters! If so, the input they provide might prove to be at best useless, and at worst, dysfunctional. Further, the customer might not have a culture that embraces close collaboration with suppliers. The culture might be one where arm's length relationships are the norm, and the idea of expending resources to provide a supplier with a demand forecast might be outside the company's normal operating procedures. Finally, even though this customer might be highly important to the supplier, and the supplier would benefit from receiving a demand forecast, the customer might not feel that the supplier is an important enough player in its supply chain to warrant the resource expenditure.

How Should This Forecasting Customer Collaboration Take Place?

The mechanics of how customer-generated forecasts can be provided range from simple and relatively informal to highly structured and highly formalized. At the most informal end of the spectrum is a simple exchange of spreadsheets on a regular basis, with key product forecasts provided on those spreadsheets. As the next section discusses, the alignment of forecasting level can be an issue. Is the detail expressed in the customer-provided spreadsheet match up with level of detail that the demand forecaster needs? Once again, the amount of data translation required is part of the cost-benefit analysis that needs to take place. A more formal and structured approach to forecasting collaboration might involve actual face-to-face meetings between the demand forecasters and the appropriate personnel at the customer's location. These appropriate personnel could be from supplier relations, procurement, material planning, or merchandising. Such meetings can take place on a regular basis—often monthly or quarterly, and the agendas can include the exchange of forecasts or other relevant business plans and to discuss specific problems or opportunities. I've worked with several companies who have regularly scheduled conference calls, accompanied by data exchanges, with their key customers, to review the numbers that are exchanged and talk through any anticipated issues.

At the far end of the spectrum in terms of formality and structure are arrangements such as Vendor Managed Inventory (VMI), or Co-Managed Inventory (CMI) or even Collaborative Planning, Forecasting, and Replenishment (CPFR[1]) agreements. Any of these arrangements require a substantial investment of resources, both human and technical by both parties—supplier and customer—from which significant benefit can be gained. Because of the substantial investment, most companies choose their formal partners carefully, investing resources such as these only in relationships where the benefit will exceed the cost, and where the level of trust between supply chain partners is strong enough to support this level of collaboration.

---

[1] Readers who are interested in learning more about CPFR should connect with the Voluntary Interindustry Commerce Solutions Association, the group that has pioneered the CPFR protocol and worked to foster collaboration and effectiveness across the supply chain, particularly in the CPG/retailer environment. For more information, visit http://www.vics.org.

How Should the Customer-Generated Forecasts Be Incorporated into Our Process?

One important element to consider with customer-generated forecasts is that these forecasts are *one of several inputs to the forecasting process.* Figure 5-2 is a graphical representation of how multiple inputs ultimately become a demand forecast. Although this looks quite simple and compelling in the figure, in reality, the details can get quite messy. As I mentioned in the previous section, depending on the level of granularity at which the customer forecast is expressed, significant translation and manipulation often must be applied to the customer's demand forecast before it can be in a form that is useful for the supplier's demand forecast. In many cases, the amount of translation and manipulation is simply not worth the effort.

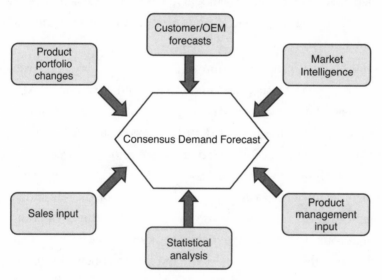

Figure 5-2 Inputs to the demand forecasting process

Summary of Customer-Generated Forecasts

Your company should consider using customer-generated forecasts. But before you do, here are four questions that those who are responsible for managing the demand forecasting process should ponder:

1. **Does our company want to get forecasts from our customers?** The answer to this question requires an analysis of the costs versus the benefits. Such an analysis frequently results in an answer of "yes," because customers can provide extremely helpful insights into their demand patterns.

2. **Assuming the answer to the preceding question is "yes," then which customers do we want to get forecasts from?** Here, the forecaster, along with the sales teams that work with these customers, needs to apply some sort of criteria to the selection of customers from whom to solicit forecasts. Strategic importance to the firm, the customer's forecasting skill, and the customer's willingness to expend the necessary resources are some of the criteria to consider.

3. **After we've decided which customers to consider for customer-generated forecasting, how should we be collaborating with these customers?** A spectrum of collaboration mechanisms exists, ranging from simple exchange of spreadsheets, to monthly or quarterly visits, to formal VMI or CPFR relationships. The maxim to keep in mind is, "The closer the collaboration, the higher the cost, but (potentially) the higher the reward."

4. **After we've worked all the preceding issues out, how do we incorporate the customers' forecasts into our process?** The key issues that need to be faced here revolve around adapting the forecasting tools to facilitative the inclusion of these forecasts, and developing a mechanism for evaluating the usefulness of the customer's forecast, relative to other sources of input.

Putting It All Together into a Final Forecast

At this point, the forecaster is faced with the challenge of data consolidation and interpretation. Chapter 4 discussed how qualitative judgments that often come from sales and marketing can contribute to the richness of the forecast. This chapter has discussed compilation of market intelligence that helps to frame the assumptions that underlie the forecast, as well as a specific form of market intelligence—customer-generated forecasts. Now it's time to turn to the task faced by demand planners of putting it all together in a way that creates a final forecast that can be discussed at the demand review.

Following the compilation of data from the various sources represented in Figure 5-2, the demand forecaster must make judgments about how to create a "final forecast" from the provided sources. Table 5-2 shows how this might look to the demand forecaster. In this simulated forecast, the analyst's job is to construct the final forecast for May through December, and actual demand is available for January through April. The first four rows in the table correspond to the categories of inputs illustrated in Figure 5-2. To populate this table, I generated random numbers in the "baseline statistical" row. Then, I randomly assigned an adjustment to the forecast for each of the rows labeled "product manager," "sales," and "customer." This procedure might be followed in a company where each of the input sources is provided with the baseline forecast, then is asked to enter an adjustment if that source felt that an adjustment was necessary. Notice that some of the cells are blank. For example, in June, the product manager had no information that would lead him or her to believe that any adjustment to the baseline statistical forecast would be necessary. Recall from Chapter 4 the discussion of how to maximize value from salesforce forecasting. I made the point there that the optimal approach for salespeople is to begin with a statistically generated forecast, and then adjust it when they have reason to believe that an adjustment will make the forecast better. The reader might want to look back at Figure 4-1, which gives guidelines for when to make such adjustments. Although the discussion in Chapter 4 referred to salespeople, it applies to any individual who is asked to provide qualitative adjustments to demand forecasts.

Table 5-2 Example of Creating a Final Forecast, Using Various Inputs

	Jan	Feb	Mar	Apr	May	Jun	Jul	Aug	Sep	Oct	Nov	Dec
Baseline Statistical	290	277	333	360	379	398	382	427	492	485	563	601
Product manager	–30	100	75		100		150		–30	–50		200
Sales		100	20			75		25		50	75	100
Customer	–40	50		75	–20		100	–50			60	40
Final forecast	255	360	381	435	419	473	507	415	462	485	631	714
Actual	290	289	313	363								
Baseline PE	0.0%	–4.2%	6.4%	–0.8%								
Product Mgr. PE	–10.3%	30.4%	30.4%	–0.8%								
Sales PE	0.0%	30.4%	12.8%	–0.8%								
Customer PE	–13.8%	13.1%	6.4%	19.8%								
Final Forecast PE	–12.1%	24.7%	21.6%	19.8%								

Returning to the example in Table 5-2, I took all judgment away from the analyst, and calculated a "final forecast" which consists of the "baseline statistical" forecast, plus (or minus) the average of the three adjustments. I also randomly generated some numbers for January through April that I labeled "actual," meaning that this was the actual demand experienced in these months. Then, to provide the analyst with a means to judge the quality of the forecasts, I calculated the percent error (PE) for each of the individual components: baseline statistical, product manager adjustment, sales adjustment, and customer adjustment. Let's examine how each source of forecast information has done. Granted, we only have 4 months of data to go by, so in "real life" we wouldn't put much credence in these results, but examining the thought process that a forecast analyst might follow is nevertheless useful.

Interestingly, all three sources of forecast adjustments made the forecast worse. In this example, the most accurate forecasts came from the baseline statistical forecast. Had this been actual (and not randomly generated) demand and forecast data, and if the analyst had considerably more than 4 months of actual performance data to work with, then that analyst would have useful insight to allow him or her to be less mechanical, and more analytical, in the creation of the final forecast. The analyst would now be able to make good business judgments about which source of forecast information was the most helpful, and identify some biases that are present in each of the various sources. (Chapter 6 covers more on measurement of bias and accuracy.)

An example of best practice is at a company that participated in the audit research several years ago. At this company, the forecasting team had gathered accuracy data for several years on the various sources of input to their forecasts. Similarly to my simulated example, this company regularly received input from its sales teams, product management teams, and several of its large customers. Also similarly, this company had calculated the extent to which each source of information made the forecast better or worse. In other words, over time, did the sales team improve upon the baseline statistical forecast? If so, by how much? What was the contribution of the product management team? Which customers improved the forecast, and which made it worse? With this data in hand, this company was able to create an overall "weighting" that was applied to each adjustment. In my simple example, I created the final forecast by adding the average of all the adjustments to the baseline forecast. This company, rather than creating a final forecast with a *simple* average of the additional inputs, created a final forecast with a *weighted* average of those inputs, with the weighting being created using historical measures of forecast accuracy. This accomplished two things: It resulted in better final forecasts, and provided incentives to the various contributors to the final forecast to make their input as helpful as it could be.

Summary

At this stage in the discussion all the relevant pieces of the demand forecast have been brought together. You've learned about the role that quantitative, or statistical, forecasting plays in the process, by looking "in the rear-view mirror" at historical demand. You've explored the important contribution that qualitative judgment makes in determining how the future is likely to look different from the past. You've also looked at the bigger picture of market intelligence, and how such market intelligence is necessary to frame the assumptions that underlie the forecast. The next piece of the puzzle returns the discussion to the adage first quoted in Chapter 2: "What gets measured gets rewarded, and what gets rewarded gets done." The next chapter provides an in-depth look at forecasting performance measurement.

6

Performance Measurement

In Chapter 2, "Demand Forecasting as a Management Process," I introduced the phrase, "What gets measured gets rewarded, and what gets rewarded gets done." Although I would like to claim 100% credit for this sublime piece of wisdom, I cannot, in good conscience. I first heard it from my colleague Tom Mentzer, and we used it extensively in our 2004 book.[1] This phrase complements, and expands upon, the phrase attributed to Peter Drucker, or maybe Edwards Deming that I also quoted in Chapter 2: "If you can't measure it, you can't manage it." I made the point then that because demand forecasting is a management process, like any other, it cannot be properly managed if it cannot be properly measured. For this reason, a book on how to best manage the demand forecasting process would be incomplete without a discussion of performance measurement.

This chapter first answers the question, "Why bother measuring forecasting performance?" Following that discussion, the chapter looks at the difference between process metrics and outcome metrics, and then explores how demand forecasting process metrics can affect outcome metrics. Next, the focus turns to the relevant demand forecasting process metrics, namely bias and accuracy. The chapter goes beyond how to calculate these metrics, and moves to the question of, "What do you do with these measures after you calculate them?" The chapter concludes with a detailed discussion of outcome metrics, and

---

[1] Mentzer, John T. and Mark A. Moon (2004), *Sales Forecasting Management: A Demand Management Approach*, Thousand Oaks, CA: Sage Publications.

gives attention to how the process metrics of accuracy and bias can be translated into the outcome metrics that ultimately drive shareholder value.

Why Bother Measuring Forecasting Performance

Chapter 7, "World-Class Demand Forecasting," covers characteristics of world-class performance in demand forecasting at length. One of the dimensions of forecasting management covered in Chapter 7 is labeled "Performance Measurement." Worst-in-class companies, labeled in Chapter 7 as "Stage 1," don't measure demand forecasting performance. A number of companies that have participated in our audit research simply do not measure forecasting performance. When interviewing individuals at these companies, we ask simple questions like "How accurate are your forecasts?" and we get answers such as, "Well, I'm not sure, but I know they could be better." Or we get answers like "Well, I think it's around 75%—something like that." At these worst-in-class companies, managers have not been convinced that measuring performance is worth the trouble. Doing it is a pain in the neck—there is more data to collect, analyze, and distribute, which takes time away from the limited time available to do forecasts. Measuring can get very complicated—do you measure at the SKU-level? The product family level? The customer level? What constitutes "good performance?" How accurate should the forecasts be? These Stage 1 companies have not yet come to grips with the "WIIFM" —What's In It For Me—question. What's In It For Me to go to all the trouble to measure performance?

My response to these companies is that a multitude of reasons exist to bother with measuring performance—pain in the neck or not! These reasons include

- Unless you measure performance, you have no way of knowing whether you're getting better or worse. The best-performing companies have cultures that encourage continuous improvement. Without a baseline measurement, followed by regular measurements against that baseline, you have no way to know whether a process is resulting in improvement or not. Thus, measures that "keep score" are important to promote this culture of continuous improvement.

- Without measures of performance, diagnosing problems with forecasting performance is impossible. Some forecasting techniques are not appropriate for the demand patterns that they are trying to predict, and performance measurement can greatly assist with this type of diagnosis.

- As discussed in Chapter 1, "Demand/Supply Integration," inventory management is a continuous "balancing act," with inventory managers doing their best to have enough inventory on hand to compensate for variability of demand, thus providing high levels of customer service, while at the same time keeping costs low by keeping inventories as low as possible. Keeping inventories at optimal levels without knowing the variability of demand is difficult, and accuracy metrics are frequently used as surrogate measures for variability of demand.

- Chapter 4, "Qualitative Forecasting Techniques," discussed the need to reward individuals for good performance in forecasting. Although I acknowledge this is oversimplifying human nature, people are all like Pavlov's dogs—we respond to stimuli! The stimuli that people tend to respond to can often be found in their performance plans, and companies that place excellence in forecasting in the performance plans of salespeople, product managers, and marketing people see improvements in their forecasting performance. However, a necessary component of creating a reward structure for forecasting performance is a measure of that performance.

For these reasons, for companies to expend resources measuring forecasting performance is indeed worthwhile. Yes, it is time consuming and requires both system and human resources, and yes, the measurements can get complicated and several tricky issues must be resolved, which are addressed in this chapter.

Process Metrics Versus Outcome Metrics

Before diving into questions of how to measure demand forecasting performance, distinguishing between process effectiveness and outcome effectiveness, and how each should be measured is useful. Examples of outcome metrics include expenditures on expedited freight, inventory turns, customer fill rates, and OTTR (On Time To Request) percentage. Companies use these important supply chain and customer service metrics to measure their performance. They can be measured either in comparison to historical performance, or against some pre-determined goal. Those are outcome metrics, and companies hope to improve these outcomes through the design and implementation of effective processes. As discussed in Chapter 1, demand/supply integration, or DSI, is a "super-process" that companies implement in the hopes of achieving important outcome metrics. This is the *raison d'etre* for DSI—to help a company achieve the outcome metrics that truly drive shareholder value, such as inventory turns, out-of-stock rates, working capital levels, and customer satisfaction. Process metrics are only interesting and important if they eventually improve outcome metrics. This chapter primarily discusses demand forecasting (a subprocess of DSI) metrics, and these forecasting metrics are good examples of process metrics. Improving forecast accuracy is a good thing, but it is only a good thing if it leads to improvement in one or more important outcome metrics. As stated in Chapter 2, no one buys stock in your company because you are good at forecasting, nor will anyone buy stock in your company based on how good you are at DSI. They will buy stock in your company because you manage your assets well, keep costs under control, and serve your customers well. That's why a *process* like DSI is so important. If DSI is done well, it translates process excellence in things such as demand forecasting into good business decisions that lead to excellence in outcome metrics such as inventory turns, costs, and fill rates.

From the perspective of the overall DSI process, Table 6-1 offers some suggestions of how to think about process effectiveness, and how your company can measure this "super-process." With this discussion of process versus outcome metrics as a backdrop, we now turn to a more detailed discussion of how to perform process measurement on the subprocess that constitutes the focus of this book: demand forecasting.

Table 6-1 Measuring DSI Effectiveness

Indicators of Effectiveness in DSI	How Would You Measure It?
Cross functional participation (horizontal effectiveness)	Are all critical business functions represented at the meetings?
	Are demand-side, supply-side, and financial personnel attending meetings when they are scheduled to attend them?
Multi-level participation (vertical effectiveness)	Are there multiple levels of meetings that involve multiple levels of participants?
	Demand review?
	Supply review?
	Reconciliation?
	Executive DSI?
Decision-oriented	Are individuals who can make gap-closing decisions in attendance at the meetings?
	Are future-oriented demand and supply forecasts the primary discussion points, or do meetings consist of "post-mortems" for why targets were not reached in the previous month?

Table 6-1 continued

Indicators of Effectiveness in DSI	How Would You Measure It?
Strategic focus	What is the relevant time horizon for forecasts, and thus plans?
	Are meetings focused exclusively on short-term issues or are longer-term demand projections and supply capabilities discussed?
Accountability	Are individuals held accountable for both participation in the process, and process-oriented metrics such as forecast accuracy?
Wide-spread training	Are individuals across the enterprise (demand, supply, and finance) participating in training to understand their roles in the process, and to learn the skills they need to effectively contribute?
Continuous process review and improvement	Is there a process in place to assess compliance with DSI?
	Is there accountability for organizations that fail to reach DSI compliance?
	Are there identified "champions" in place who act as process owners and who drive continuous improvement?

Measuring Forecasting Performance

The most common, and useful, categories of forecasting performance that you can measure are *accuracy* and *bias*. You can think of accuracy as the difference between forecasted demand and actual demand. You can think of bias as systematic patterns of either forecasting too high or forecasting too low.

The discussion that follows focuses only on those measurements that are most commonly used in practice. Because this book is designed to be a guide for business professionals on how to best manage their demand forecasting processes, I do not provide a comprehensive discussion of all possible ways to measure either accuracy or bias.[2] Rather, I focus on the most commonly used. Both bias and accuracy metrics start with the foundational calculation of *percent error,* or *PE.* The following sections first discuss the calculation of PE and some of the issues surrounding that calculation, then moves on to describe how PE is used to uncover bias in a forecast, and how it is used as a scorecard metric to track accuracy.

The Building Block: Percent Error

The most commonly used tools to identify bias and measure accuracy are built upon the simple calculation of percent error. Percent error tells you, in relative terms (that is, it's expressed as a percentage, not as an absolute number), how far off the forecast was from actual demand, and whether the forecast was too high or too low. The formula for percent error is

$$Percent\ Error = PE = \frac{(Forecasted\ Demand - Actual\ Demand)}{Actual\ Demand} * 100$$

Several issues arise from this formula. One is the question of what belongs in the denominator: forecast or actual? Various books and articles expound on the proper calculation. Some authors argue that actual demand must be in the denominator of the equation, as is expressed in the preceding. Others argue that forecasted demand should be in the denominator. From a philosophical perspective, I favor putting actual demand in the denominator, because then the calculation describes how far off the forecast was, relative to *what actually happened.* When the denominator contains forecasted demand, then percent error describes how far off the forecast was, relative to *what you guessed was going to happen.* The results from using these different formulas can be quite dramatic, as illustrated in Table 6-2, which contains randomly generated numbers in the "Actual Demand"

---

[2] For those readers who would like a comprehensive discussion of multiple ways to measure forecast accuracy and bias, I highly recommend Mentzer, John T. and Mark A. Moon (2004), *Sales Forecasting Management: A Demand Management Approach,* Thousand Oaks, CA: Sage Publications. Chapter 6 in this volume contains a detailed discussion of many different ways to measure accuracy and bias.

column and the "Forecasted Demand" column. In both cases, the random numbers are between 500 and 1,500. If the forecaster were to choose to calculate percent error with actual demand in the denominator, he or she would conclude that the average percent error over the 24 months shown in the Table is –25%, meaning that on average, the forecast is 25% lower than actual demand. On the other hand, if that same forecaster were to choose to put forecasted demand in the denominator, then the conclusion would be that the average percent error over the preceding 24 months was –4%. Quite a difference!

Table 6-2 Comparison of Percent Error Calculation: "Actual" in Denominator versus "Forecast" in Denominator

	Actual Demand	Forecasted Demand	PE with Actual in Denominator	PE with Forecast in Denominator
Jan-09	1066	540	–97.41%	–49.34%
Feb-09	572	645	11.32	12.76
Mar-09	1291	994	–29.88	–23.01
Apr-09	923	1159	20.36	25.57
May-09	994	754	–31.83	–24.14
Jun-09	1139	762	–49.48	–33.10
Jul-09	882	524	–68.32	–40.59
Aug-09	776	739	–5.01	–4.77
Sep-09	1485	523	–183.94	–64.78
Oct-09	1087	1043	–4.22	–4.05
Nov-09	846	863	1.97	2.01
Dec-09	924	765	–20.78	–17.21
Jan-10	1168	815	–43.31	–30.22
Feb-10	1304	1316	0.91	0.92
Mar-10	1131	710	–59.30	–37.22
Apr-10	1431	901	–58.82	–37.04
May-10	1301	651	–99.85	–49.96
Jun-10	1234	1239	0.40	0.41
Jul-10	526	1104	52.36	109.89
Aug-10	1226	1013	–21.03	–17.37
Sep-10	1014	899	–12.79	–11.34
Oct-10	1125	1262	10.86	12.18

	Actual Demand	Forecasted Demand	PE with Actual in Denominator	PE with Forecast in Denominator
Nov-10	972	1232	21.10	26.75
Dec-10	575	1463	60.70	154.43
Average Percent Error over 2 years of data			–25.25	–4.13

At this point, you should consider the question, "Does it really matter?" The answer is, "yes and no." On the one hand, if what you need is an absolute measure of accuracy, then it does matter. On the other hand, if what you need is to compare performance between periods, or between products, or between customers, or between salespeople, then it matters less, because as long as it's measured the same way each period, you can make rational comparisons.

Let's reflect on what you want to accomplish by measuring performance. The four primary goals of measuring performance are

1. Tracking to see whether the forecasting process is getting better or worse
2. Diagnosing problems with specific forecasting techniques or tools
3. Measuring demand volatility to aid in inventory decisions
4. Tracking individual performance to facilitate rewarding forecasting excellence

For goals number 1 and 4, it shouldn't matter if you use forecast or actual in the denominator. As long as the metric is calculated the same each time, then the forecaster can still use the calculation to make comparisons between time periods, individual forecasters, customers, and so on. However, for goals number 2 and 3, it does matter which term is in the denominator. If the inventory planner uses forecast accuracy as a surrogate of demand volatility, is average demand volatility in the example 25%? Or 4%? Obviously, that matters. Similarly, if the forecaster is examining the performance metric as a way to diagnose problems, then the magnitude of the calculated error can be significantly different, depending on which term is in the denominator, and this could easily lead to a misdiagnosis. The bottom line is that because it does matter which term is in the denominator for some of the purposes that forecasting performance measurement is designed to accomplish, then choosing the calculation that is the most satisfying, which is, in this author's opinion, to have actual demand in the denominator, benefits the forecaster.

A second issue that arises from this formula lies in the "forecasted demand" number. If forecasting takes place monthly, and if the forecast is a rolling 18–24 month forecast, then the question becomes how many months prior to the month being forecasted should the forecast be "locked" for purposes of calculating accuracy? For example, in Table 7-2, the January 2009 forecast is 540 units. Is that the number that was forecasted for January when the forecast was completed in July 2008? Or December 2008? One would hope that the forecast for January demand that was completed in December would be more accurate than the forecast for January demand that was completed in July, because in December, much more up-to-date information is available about what demand is likely to be. You can derive the answer to this question using the same logic used in Chapter 2 during the discussion about the appropriate time horizon for forecasting. Recall from that discussion that a forecasting horizon should be *at least as long as the lead time for producing the product being forecasted.* In other words, if a product's lead time is 3 months from the time it is ordered until the time it is delivered to the customer, then a forecast horizon of 2 months is not particularly useful. The same logic applies here. The forecaster should "lock in" the forecast consistent with the product's lead time and calculate the percent error using that period's forecasted demand as the "forecast" term in the equation.

The final issue that's worthy of discussion is the term *actual* in the equation for percent error. Here, the issue is one of "what was actually demanded by the customer" versus "what was shipped to the customer." As discussed in Chapter 7, and previously discussed in Chapter 2, often the case is that the quantity demanded by customers is different from the quantity shipped to customers. Forecasters are urged to forecast *true demand*, meaning *what would customers buy from us if they could*? If that is the basis of the *forecast*, then that needs to be the basis of the *actual*. In other words, having the *actual* term in the percent error calculation represent what was *actually demanded* during the time period being measured, not what was *actually shipped* is critical. This book discusses how to best capture this actual demand, but comparing forecasted *demand* with actual *sales* is not correct. Doing that is the proverbial "comparing apples with oranges."

Identifying Bias

You can best think of the first performance measurement, *bias*, as a systematic pattern of overforecasting (upward bias) or underforecasting (downward bias). The easiest way to reveal bias is through examination of percent error, as represented graphically. Take a look at a couple of PE charts to examine some of the issues involved in identifying bias.

Figure 6-1 shows the first example. Let's begin examining this chart by describing what it shows. The vertical axis in the chart represents the level of percent error that has been calculated over the preceding 24 months. A data point of 0 would indicate that the forecasted demand and actual demand were identical. As expected, the chart has no data points of 0, consistent with my statement in Chapter 1 that one truth about forecasts is that they are always, always *wrong!* A data point that is a positive number indicates that the forecast was higher than actual demand—an overforecast—and a negative data point indicates that the forecast was lower than actual demand—or an underforecast. Typically, when an overforecast takes place, the result is building inventory, and when an underforecast takes place, the result is drawing down of inventory, or problems with satisfying customer demand.

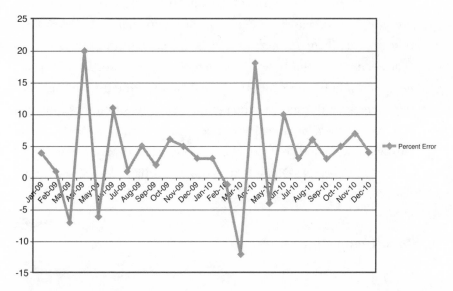

Figure 6-1 Identifying bias—Example 1

Now, let's examine the patterns that are displayed in Figure 6-1. In January and February of both years, you can see a modest over-forecast. In March, the percent error dips negative, and then in April, it shoots highly positive. In May, it returns to an underforecast, June goes back to overforecast, then settles in to a modest overforecast from July through the next January. So what in the world is going on here?

As a forecast analyst, you are presented here with a wealth of infor-mation. First, the obvious issue to draw your attention is the spike of underforecasting in March, followed by a reaction of overforecasting in April, then under in May and over in June. The PE chart in Figure 6-1 doesn't tell you what's going on here, but it does point out some sort of repeating pattern that deserves further investigation. A couple hypotheses could be offered. One is that a particular customer might have an ordering pattern that's in a different time window than you thought. You forecasted their spike in demand to come in a different month than what actually is occurring. Another possibility is that when a relative minor underforecast occurred in March, an overreaction occurred on the part of the demand forecaster, who overcompensated in April, then overcompensated again in May, and it wasn't until July that the steady-state returned. Just by looking at the chart, you don't know which of these, or some other, explanations are correct. But you do know that a problem exists in the forecast for this product that takes place repeatedly in the March through June timeframe, and this shows the diagnostic value of examining percent error charts.

A second issue that you can address through examination of the PE chart in Figure 6-1 is that question of whether or not this is a biased forecast. Recall the definition of *bias*—systematic pattern of over- or underforecasting. Examination of Figure 6-1 should lead to the conclusion that this is indeed a biased forecast, with the bias being in the direction of overforecasting. Apart from the anomalies discussed earlier between March and June, the forecasted demand is always higher than the actual demand, indicating bias. Of course, as a business manager, one must ask the question, "Is this biased forecast problematic?" The answer to that question again is, "It depends." Let's look at the extent of bias. Apart from the anomalies that occur between March and June, the extent of bias is less than 5% positive bias. On the surface, that seems to be of little concern—most forecasters would turn cartwheels with 5% forecast error! But before you become too pleased with your performance, look deeper. For example, you don't know whether the forecast error you are examining is at the SKU level, the product family level, the brand level, or the company level. If you have 5% positive bias at the SKU level, you might indeed be able to turn cartwheels. But if you have 5% positive bias at the company level, you might be building inventory to highly unacceptable levels. Again, it depends.

Another issue is the dollar value of the product being forecasted. An example can help illustrate. Consider a company such as Honeywell. Honeywell has a large division called Honeywell Aerospace, which is a major supplier of the commercial aircraft and defense industries. They manufacture an enormous range of products that support these industries—everything from jet engines that sell for millions of dollars to the nuts and bolts that hold components in place that might be valued at less than a dollar. Obviously, if Honeywell Aerospace were able to forecast demand for "nuts and bolts" with a 5% positive bias, cartwheels would ensue. However, the case might be that if they forecasted jet engines with a 5% positive bias, the value of the excess inventory would be financially catastrophic. So, it depends.

Finally, the extent to which bias is problematic depends on strategic considerations as well. These strategic considerations address the question, "What is the cost to our firm of holding inventory versus being out of stock." One company in our audit database is a relatively small, family-owned company that sold its products through big-box home improvement stores such as Home Depot and Lowe's. Over half of its revenue came from these large retailers, who maintain very high standards for on-time delivery and fill rates. The company we worked with would, for all intents and purposes, be out of business if it failed to meet these requirements from these large customers. At this company, then, carrying some excess inventory was a much smarter strategic direction than risking being out of stock. For it, a slightly upward bias in the forecast would be considered highly acceptable.

Let's examine one more example, shown in Figure 6-2, of a PE chart to see what it can reveal about bias. Again, the vertical axis on this graph is percent error, with 0 representing a perfectly accurate forecast (forecasted demand = actual demand). January 2009 shows about an 18% underforecast (that is, the actual demand was about 18% less than the forecasted demand). What you see in this example appears to be a seasonal pattern of forecast error. Overforecasting occurs in April through October, while underforecasting occurs in November through March. What conclusions can you draw here?

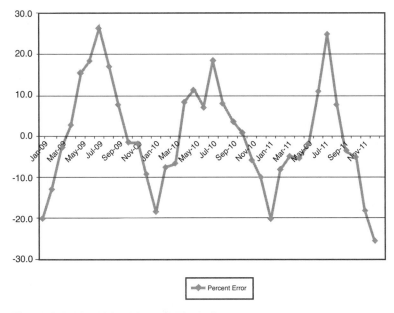

Figure 6-2 Identifying bias—Example 2

One thing you can conclude is that in total, no bias exists in this forecast. If you look only at the segment of data between April and October 2009, you can conclude that the forecast carries a positive bias. If you look only at November 2009 through March 2010, you can conclude that the forecast carries a negative bias. However, the entire stream of data, over 3 years, should not lead to a conclusion that bias exists. Bias implies that something about the process *systematically* results in over- or underforecasting caused either by qualitative judgment of people or characteristic of the quantitative forecast. Previous chapters have examined some of these sources of bias. For example, if average is used as the time-series statistical tool, and an upward trend in demand exists, then the forecast will be biased low. Also remember how qualitative judgments can result in systematic bias. If salespeople perceive that their forecasts will influence their quotas, then their forecasts might be systematically biased low. However, none of these systematic sources of bias appears to be present in the forecasting process depicted in Figure 6-2. So although the forecasting results shown in Figure 6-2 are not biased, they are certainly not very good! Thus, there is more that the PE chart can reveal. The repeating pattern of over- and underforecasting suggests two possible problems: Either the underlying pattern of demand is seasonal, but the forecasting technique does not take seasonality into account, or the forecast is predicting seasonality, but no seasonality actually is occurring in the demand. Again, examination of this PE chart serves a diagnostic purpose. It might not tell us exactly what is going on, but it does suggest verifiable hypotheses that you can further explore, which will ultimately make the forecasting process better.

Measuring Accuracy

Bias is one process metric that relates to the demand forecasting process, and the most commonly used way to document bias is through examination of percent error charts. The discussion now moves to measuring accuracy, again beginning with percent error as the foundational calculation. Accuracy is most commonly measured using an average of percent error over a certain number of historical periods. However, simply averaging percent error causes some problems. To overcome these problems, consider a new metric: Mean Absolute Percent Error, or MAPE. MAPE expresses, on average and in relative terms (that is, it's a percentage), how inaccurate the forecast has been historically. The formula for MAPE is

Mean Absolute Percent Error = MAPE = $\Sigma\,|PE|\,/\,N$

where:

N = number of periods for which the errors have been tracked

|PE| = the absolute value of the percent errors from the previous periods (that is, drop the negative signs)

The problem that MAPE overcomes is that when an average is taken of negative and positive numbers, the average will be lower than it should be because the positive numbers will be offset by the negative numbers. As an extreme illustration, consider 2 months of data where the percent error in month 1 is 90%, and the percent error in month 2 is –90%. If you take the average of these two months, then it will appear that overall forecast error is 0! Obviously, that's not a useful metric. Revisit the data shown earlier in Table 6-1, with an additional column added, to illustrate this problem (see Table 6-2), and how MAPE overcomes it.

The example in Table 6-2 illustrates why the additional step of calculating the absolute value of the percent error, when calculating accuracy, is critical. If your goal is to determine how accurate the forecast has been, you need to know the absolute deviation of the forecasted demand from the actual demand. In the example in Table 6-1, the true forecast error is 40.25%, and not –25.25%. Although MAPE is the most commonly used metric of forecast accuracy, some companies prefer to take a "glass half full" rather than "glass half empty" approach, and calculate Mean Absolute Percent Accuracy instead. Mean Absolute Percent Accuracy is simply 1 – MAPE.

Because MAPE is the more commonly used terminology, the following discussion focuses on it.

To see how useful MAPE is, revisit the reasons for measuring forecasting performance in the first place. The four primary goals of measuring performance are

1. Tracking to see whether the forecasting process is getting better or worse

2. Diagnosing problems with specific forecasting techniques or tools

3. Measuring demand volatility to aid in inventory decisions

4. Tracking individual performance to facilitate rewarding forecasting excellence

Table 6-2 Calculation of Mean Absolute Percent Error

	Actual Demand	Forecasted Demand	Percent Error	Absolute Value of Percent Error
Jan-09	1066	540	–97.41	97.41
Feb-09	572	645	11.32	11.32
Mar-09	1291	994	–29.88	29.88
Apr-09	923	1159	20.36	20.36
May-09	994	754	–31.83	31.83
Jun-09	1139	762	–49.48	49.48
Jul-09	882	524	–68.32	68.32
Aug-09	776	739	–5.01	5.01
Sep-09	1485	523	–183.94	183.94
Oct-09	1087	1043	–4.22	4.22
Nov-09	846	863	1.97	1.97
Dec-09	924	765	–20.78	20.78
Jan-10	1168	815	–43.31	43.31
Feb-10	1304	1316	0.91	0.91
Mar-10	1131	710	–59.30	59.30
Apr-10	1431	901	–58.82	58.82
May-10	1301	651	–99.85	99.85
Jun-10	1234	1239	0.40	0.40
Jul-10	526	1104	52.36	52.36
Aug-10	1226	1013	–21.03	21.03
Sep-10	1014	899	–12.79	12.79
Oct-10	1125	1262	10.86	10.86
Nov-10	972	1232	21.10	21.10
Dec-10	575	1463	60.70	60.70
			Mean Percent Error: –25.25	**Mean Absolute Percent Error (MAPE) 40.25**

The examination of PE charts is the best way to achieve goal #2—diagnosing problems with specific forecasting techniques or tools. But percent error on its own will not help with the other three performance measurement goals. For those, you need to examine percent error over a range of forecast periods. If you want to measure the overall process outcome, and see whether the process is getting better or worse, MAPE is a good "scorecard" metric to help you do that. As an overall measure of demand volatility for a particular product, or customer, MAPE is a good measure. Finally, MAPE is a good way to track the overall forecasting performance of individuals, because it documents how close their efforts came, in absolute terms, to a perfect forecast. Thus, for three of the four goals of measuring forecasting performance, MAPE is the right tool for the job.

The Advantages of MAPE

Like any tool, MAPE has its advantages and its problem. Let's initially examine MAPE's advantages:

- The use of the absolute value sign in the equation ensures that over- and underforecasts won't cancel each other out. This results in a more complete assessment of forecasting effectiveness over time.

- You can use MAPE to "keep score" for whatever level of the forecasting hierarchy that the manager needs. As long as a complete forecasting hierarchy is in place (refer to Chapter 2 for the discussion of forecasting hierarchy), then accuracy can be measured at the SKU, product-line, brand, customer, location, or whatever other level is needed.

- MAPE can be weighted to account for differences between products. An example can help illustrate the usefulness of

MAPE in this regard. Suppose you wanted to measure the forecast accuracy for a product family that consists of two products: nuts and bolts, and turbine blades for jet engines. The nuts and bolts product is characterized by high volume and high volatility, but the unit price is low at $10 per unit. Turbine blades, on the other hands, can be characterized as low volume, low volatility, and high unit price of $10,000 per unit. Table 6-3 shows the data for this hypothetical example. This table contains one year of actual demand for both nuts and bolts, and turbine blades, along with the calculation of percent error, absolute percent error, and MAPE for each individual product. As you would expect, MAPE for the highly volatile nuts and bolts is quite high, at 42.4%, while MAPE for the low-volatility turbine blades is a respectable 2.24%.

Table 6-3 Calculation of Weighted MAPE—Raw Data

Nuts and Bolts

	Actual Demand	Forecasted Demand	Percent Error	Absolute Value of Percent Error
Jan-11	42216	40949	3.09%	3.09%
Feb-11	46901	25024	87.42%	87.42%
Mar-11	49154	39116	25.66%	25.66%
Apr-11	36652	18450	98.66%	98.66%
May-11	31996	26273	21.78%	21.78%
Jun-11	28748	45001	-36.12%	36.12%
Jul-11	27608	15526	77.82%	77.82%
Aug-11	26182	46549	-43.75%	43.75%
Sep-11	21073	39155	-46.18%	46.18%
Oct-11	33522	43538	-23.01%	23.01%
Nov-11	21204	34047	-37.72%	37.72%
Dec-11	19038	20597	-7.57%	7.57%
			MAPE for Nuts and Bolts	42.40%

Turbine Blades

	Actual Demand	Forecasted Demand	Percent Error	Absolute Value of Percent Error
Jan-11	227	236	-3.81%	3.81%
Feb-11	244	240	1.67%	1.67%
Mar-11	244	237	2.95%	2.95%
Apr-11	237	237	0.00%	0.00%
May-11	241	236	2.12%	2.12%
Jun-11	227	225	0.89%	0.89%
Jul-11	232	236	-1.69%	1.69%
Aug-11	225	229	-1.75%	1.75%
Sep-11	227	241	-5.81%	5.81%
Oct-11	238	245	-2.86%	2.86%
Nov-11	247	241	2.49%	2.49%
Dec-11	235	237	-0.84%	0.84%
			MAPE for Turbine Blades	2.24%

Table 6-4 presents a calculation for the product family MAPE, which is not weighted in any way. The actual demand and forecasted demand are simply the sums of the demand for each of the two products in the product family, and the unweighted MAPE is a not-so-stellar 42.01%.

Table 6-4 Calculation of Unweighted Aggregate MAPE

Product Family Aggregated

	Actual Demand	Forecasted Demand	Percent Error	Absolute Value of Percent Error
Jan-11	42443	41185	3.05%	3.05%
Feb-11	47145	25264	86.61%	86.61%
Mar-11	49398	39353	25.53%	25.53%
Apr-11	36889	18687	97.40%	97.40%
May-11	32237	26509	21.61%	21.61%
Jun-11	28975	45226	−35.93%	35.93%
Jul-11	27840	15762	76.63%	76.63%
Aug-11	26407	46778	−43.55%	43.55%
Sep-11	21300	39396	−45.93%	45.93%
Oct-11	33760	43783	−22.89%	22.89%
Nov-11	21451	34288	−37.44%	37.44%
Dec-11	19273	20834	−7.49%	7.49%
Unweighted aggregate MAPE for product family			42.01%	

The question now becomes, is this MAPE of 42.01% a useful metric to judge the forecasting accuracy of this product family? The calculation is correct (at least I hope it's correct!), but is it useful? I would argue that it is not, because the forecasting performance of the turbine blades, which is the product that really matters from a financial and supply chain perspective, is quite good, while the product that matters far less—nuts and bolts—is not forecasted as accurately, but probably doesn't need to be forecasted as accurately. Fortunately, you can adapt the MAPE calculation to get a more satisfying assessment of the forecasting performance of this product family. Table 6-5a and b provide this example. Here, you utilize a *Weighted Aggregate MAPE,* the formula for which is

$$Weighted\ Aggregate\ MAPE = \sum_{p=1}^{P} MAPE_p * (D_p \div D_T)$$

where:

$MAPE_p$ = MAPE for product p

D_p = Dollar demand for product p

D_T = Total dollar demand for all P products

Table 6-5a Calculation of Weighted Aggregate MAPE

Nuts and Bolts

	Actual Demand	Forecasted Demand	Percent Error	Absolute Value of Percent Error	Total Dollar Demand, Selling Price of $10/Unit
Jan-11	42216	40949	3.09%	3.09%	$422,160.00
Feb-11	46901	25024	87.42%	87.42%	$469,010.00
Mar-11	49154	39116	25.66%	25.66%	$491,540.00
Apr-11	36652	18450	98.66%	98.66%	$336,520.00
May-11	31996	26273	21.78%	21.78%	$319,960.00
Jun-11	28748	45001	−36.12%	36.12%	$287,480.00
Jul-11	27608	15526	77.82%	77.82%	$276,080.00
Aug-11	26182	46549	−43.75%	43.75%	$261,820.00
Sep-11	21073	39155	−46.18%	46.18%	$210,730.00
Oct-11	33522	43538	−23.01%	23.01%	$335,220.00
Nov-11	21204	34047	−37.72%	37.72%	$212,040.00
Dec-11	19038	20597	−7.57%	7.57%	$190,380.00
					Annual total
			MAPE	42.40%	$3,842,940.00

Weighting is calculated by dividing the annual total dollar value of nuts and bolts by the annual total dollar value of the product family as a whole	**Weighting**	**Weighted MAPE (MAPE × weighting)**
$3,842,940 / ($3,842,940 + $28,240,000)	**0.1198**	**5.08%**

Table 6-5b Calculation of Weighted Aggregate MAPE (continued)

Turbine Blades

	Actual Demand	Forecasted Demand	Percent Error	Absolute Value of Percent Error	Total Dollar Demand, Selling Price of $10,000/ Unit
Jan-11	227	236	–3.81%	3.81%	$2,270,000.00
Feb-11	244	240	1.67%	1.67%	$2,440,000.00
Mar-11	244	237	2.95%	2.95%	$2,440,000.00
Apr-11	237	237	0.00%	0.00%	$2,370,000.00
May-11	241	236	2.12%	2.12%	$2,410,000.00
Jun-11	227	225	0.89%	0.89%	$2,270,000.00
Jul-11	232	236	–1.69%	1.69%	$2,320,000.00
Aug-11	225	229	–1.75%	1.75%	$2,250,000.00
Sep-11	227	241	–5.81%	5.81%	$2,270,000.00
Oct-11	238	245	–2.86%	2.86%	$2,380,000.00
Nov-11	247	241	2.49%	2.49%	$2,470,000.00
Dec-11	235	237	–0.84%	0.84%	$2,350,000.00
					Annual total
			MAPE	2.24%	$28,240,000.00

Weighting is calculated by dividing the annual total dollar value of turbine blades by the annual total dollar value of the product family as a whole	**Weighting**	**Weighted MAPE (MAPE × weighting)**
$28,240,000 / ($3,842,940 + $28,240,000)	**0.8802**	**1.97%**
Weighted aggregate MAPE for product family (sum of weighted MAPE's for both products in product family)	**7.05%**	

By using this formula and working through the arithmetic, you can see in Table 6-5a and b that the weighted aggregate MAPE for the product family as a whole (two products) is 7.05%. The very good forecast for the high-dollar-value product (turbine blades) carries far more weight in the calculation than the less good forecast for the low-dollar-value product (nuts and bolts). This is a far more useful representation of the actual forecasting performance of this hypothetical product line as a whole.

The Disadvantages of Using MAPE

Now that the advantages of MAPE have been examined, the disadvantages of MAPE need to be considered as well:

- One disadvantage of using MAPE is that if too many historical periods are used in the calculation, MAPE becomes sluggish and unable to reflect recent improvements, or decrements, in forecasting performance. A common industry practice that alleviates this problem is to use a rolling MAPE of some standard length—most commonly a 12-month rolling MAPE—as the standard accuracy metric. Such an adaptation to the MAPE calculation provides a good balance between not allowing any one period—either good or bad—to overly influence the average, while at the same time not allowing too many periods to be used, which makes the metric unresponsive.

- Another common problem in using MAPE is that it can lead to unrealistic accuracy targets. A brief story might be useful to illustrate this point. Several years ago, our audit team worked with a company that was struggling with its demand forecasting. The average SKU-level MAPE was around 60%. Inventory was piled everywhere, but it was often the wrong inventory, and customer fill rates were unacceptably low. Following our audit, where we recommended extensive changes to their processes, culture, and tools, this company began to see encouraging results. After two years of commitment to re-engineering its forecasting efforts, the metrics were all moving in the right direction. Average SKU-level MAPE had improved into the upper 20% range, inventories were down dramatically, and customers noticed the improvements in their fill rates. However, our contacts in the company reported back to us that in spite of all this success, morale in the forecasting department was rock bottom, and turnover was higher than they had ever seen. Our team was invited back to re-audit the company's processes, and we found an interesting phenomenon. Senior management at this company had decided that they wanted to be very aggressive with their targets for forecasting excellence, and they gave the forecast analysts a target of SKU-level MAPE of no more than 15%. The forecasters were deflated. In spite of their hard work, their dedication, and their willingness to adapt, and in spite of the supply chain successes they had achieved, they were continually denied the bonus money that would have been available had they reached the arbitrary target of 15% MAPE. This story has several morals. One is that rewarding individuals based upon *period-over-period improvement*, rather than arbitrary accuracy targets, is far better. Some products are simply highly volatile by nature, and practical limits might exist on the level of forecast accuracy that can be achieved. Another moral is that, as I've stated repeatedly, *no one buys stock in a company because it is good at forecasting*! As discussed later in greater detail in Chapter 7, world-class companies keep their eyes on the true prizes—shareholder value, reduced inventories, improved fill rates, lower expediting costs—and recognize that forecast accuracy, though a worthy *process metric*, is only a means to an end.

Outcome Metrics—The Results of Forecasting Excellence

This chapter on performance measurement concludes by summarizing a wonderful article written by the late Tom Mentzer and published in the *Journal of Business Forecasting* in 1999.[3] This brief article beautifully illustrates the fact that improving performance in demand forecasting (that is, improving process metrics) can result in dramatically improved outcome metrics that benefit the firm as a whole. Figure 6-3, adapted from Mentzer's article, shows a simplified version of the "DuPont Model," which is an approach developed by the DuPont Corporation in the 1920s to analyze return on equity. In this analysis, return on shareholder value is a simple equation that combines the income statement (top half of Figure 6-3) with the balance sheet (bottom half of Figure 6-3). Return on shareholder value, then, is profit divided by capital invested (ignore retained earnings to make the discussion simpler). Profit (the income statement element) is revenue minus costs, and capital invested is (for purposes of this simplified analysis) working capital, which is inventory plus accounts receivable minus accounts payable. Again, to simplify the discussion, ignore the fixed capital element of capital invested.

---

[3] Mentzer, John T. (1999), "The Impact of Forecasting Improvement on Return on Shareholder Value," 18 *Journal of Business Forecasting*, (Fall), 8–12.

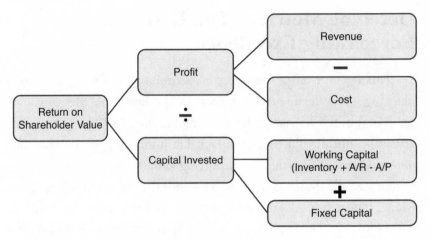

Figure 6-3 The impact of forecasting process improvement on shareholder value

Figure 6-4 illustrates the fascinating way that Mentzer used this simplified DuPont model to illustrate the value of forecasting process improvement. The additional financial detail included in Figure 6-4 comes from a company that participated in our audit research. When we audited this company, we found many challenges. It was not using any statistical analysis of historical demand to model repeating patterns. It was not effectively utilizing its sales organization or its product managers to gain qualitative insights about changing customer or product demand. It was not measuring or rewarding forecasting performance. It also had no DSI process in place taking demand forecasts and turning them into good business plans. But this company showed itself to be a great company by taking the insights it learned from the audit, and embarking on a 2-year journey to transform its demand forecasting processes. The company spent approximately $2 million on this process improvement effort. It bought an up-to-date forecasting system that helped it to statistically model historical demand. It hired more demand forecasters and trained everyone involved in forecasting on both methods and process excellence. It convinced the senior sales executive to adapt the incentive system for the sales team to include forecast accuracy as one of the components.

Also, and perhaps most critically, it named an extremely effective manager to be the "forecasting champion," and he relentlessly drove the organization toward process improvement.

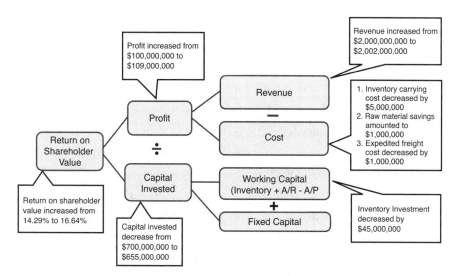

Figure 6-4 The impact of forecasting process improvement on shareholder value: an actual example.

Figure 6-4 documents the startling results.[4] The first effect was on revenue. The company documented an improvement in its out-of-stocks, and because it more often had product available to sell to customers, when customers wanted it, it increased its annual revenue by .1%. The next category of effects was cost. Because the company decreased its inventory investment substantially, its annual inventory carrying costs (which appear in the income statement) decreased by $5,000,000. An interesting benefit came in the area of raw material purchases. This company had one particular raw material that was an extremely expensive component of its production process. Because its forecasts had been so poor historically, it found itself purchasing large quantities of this raw material on the "spot market," rather than

---

[4] The actual numbers in Figure 6-4 are disguised at the request of the company involved, but the overall effect in each category is accurate.

through long-term contracts with its supplier. However, after the forecasting process improved, the purchasing department reported that they now "trusted the forecast," and were able to enter into a long-term contract with the supplier that led to significant cost savings of $1,000,000. Finally, the company documented a reduction in expediting freight of $1,000,000 annually. This came about from more consistently having the right product at the right place at the right time—when and where customers needed it. When you consider this outcome in combination with the next outcome—a reduction in the company's inventory levels of $45,000,000, it becomes clear that this company experienced some of the "DSI magic" referred to Chapter 1. This is the "magic" of improving customer service, reducing operating costs, and reducing inventory levels—*all at the same time.* When you do the arithmetic, you will see that this company improved its Return on Shareholder Value from 14.29% to 16.64%, thanks to an investment of about $2 million. That's an outcome that any CFO would be turning handsprings to see!

Summary

This chapter began with the adage, "What gets measured gets rewarded, and what gets rewarded gets done." It's an important enough adage that the chapter ends with the same advice. This chapter discussed why measuring forecasting performance (a process metric) is important, how to measure forecasting in such a way that both accuracy and bias can be documented, and how to use forecasting metrics to achieve four important results:

1. Tracking to see whether the forecasting process is getting better or worse

2. Diagnosing problems with specific forecasting techniques or tools

3. Measuring demand volatility to aid in inventory decisions

4. Tracking individual performance to facilitate rewarding forecasting excellence

This chapter also looked at the importance of keeping your eyes on the prize—shareholder value. As I've said to many audiences, "An accurate forecast and 50 cents will buy you a cup of coffee." In spite of my need to update that statement in the Age of Starbucks, the point is still true. The prize consists of outcome metrics that can be driven by the forecasting metrics discussed in this chapter.

The discussion now turns to the topic of world-class forecasting. You've learned about forecasting techniques, processes, and measurement, and now we will look carefully at what constitutes best—and worst—practices in applying these concepts in real business organizations.

7

World-Class Demand Forecasting

The efforts to understand what constitutes world-class performance in demand forecasting began, for members of the University of Tennessee research team, back in 1984 when none of the researchers involved was even a member of the University of Tennessee faculty! It was in 1984 that my colleague, Tom Mentzer, then a professor at Virginia Tech University, began his series of published papers that documented the state of forecasting practice in industry, which became known as "the Benchmark Studies." The first two papers were published in the *Journal of Forecasting*, and they documented two large-scale survey efforts that took place ten years apart, the purpose of which was to take a snapshot of current practice in forecasting. The first study was labeled "Phase 1" and was published in 1984.[1] The authors surveyed 157 companies, and documented the extent to which various forecasting techniques, both statistical and qualitative, were being utilized in practice. Now nearly 30 years old, this article reflected the focus of forecasting research at that time: namely, techniques. Most forecasting research was centered on statistical matters—devising and testing various statistical algorithms for modeling various types of demand patterns. Ten years later, a second study was undertaken, labeled as Phase 2, and published in 1995.[2] To a large

[1] Mentzer, John T. and James E. Cox, Jr. (1984), "Familiarity, Application, and Performance of Sales Forecasting Techniques," *Journal of Forecasting*, 3, 27–36.

[2] Mentzer, John T. and Kenneth B. Kahn (1995), "Forecasting Technique Familiarity, Satisfaction, Usage, and Application," *Journal of Forecasting*, 14 (No. 5), 465–476.

degree, the Phase 2 study replicated the methodology of the Phase 1 study, with a major exception. Reflecting the changing realization that forecasting excellence is much more than the selection of the right statistical algorithm, Mentzer and his colleagues also collected data from another large group of survey respondents—208 in this case—on the management approaches and systems implementations that they were using. Interestingly, even though more than 10 years had passed between the first and second studies in this series, the findings showed that in spite of advances in statistical sophistication, management focus, and computer system assistance, overall forecasting performance had not, on average, improved much at all!

This surprising finding led Mentzer and his team to embark upon a qualitative effort to understand the state of forecasting in industry. In Phase 3, the team selected 20 companies from a variety of industries and across various levels of their supply chains,[3] and then conducted in-depth face-to-face interviews with a broad spectrum of individuals from those firms who were involved with forecasting. Individuals from forecasting functions, as well as those who provided input to forecasts (such as sales and marketing) and those who were the "customers" of forecasts (such as manufacturing, production planning, procurement, and finance) were interviewed. The results of this effort were published in *Business Horizons*,[4] and this research brought tremendous clarity to the many of the unanswered questions from the first two phases of the Benchmark Studies.

By this point, Mentzer had moved to the University of Tennessee, and beginning in 1996, I became part of the research team. It was at that point that I began to use many of the insights gained in Phases 1–3

---

[3] The 20 companies that participated in the third phase of the benchmark studies were Anheuser-Busch, Becton-Dickinson, Coca-Cola, Colgate Palmolive, Federal Express, Kimberly-Clark, Lykes Pasco, Nabisco, J.C. Penney, Pillsbury, Pro-Source, Reckitt Colman, Red Lobster, RJR Tobacco, Sandoz, Schering Plough, Sysco, Tropicana, Warner Lambert, and Westwood Squibb.

[4] Mentzer, John T., Carol C. Bienstock, and Kenneth B. Kahn (1999), "Benchmarking Sales Forecasting Management," *Business Horizons*, (May–June), 48–56.

of the Benchmark Studies to begin working directly with companies, conducting what we called *forecasting audits*. Fast-forward 16 years to the time of this writing, and our team has now completed 43 audits for companies around the world, from manufacturers to retailers, from consumer packaged goods companies to heavy industry companies.[5] The methodology for conducting a forecasting audit has been well documented in the academic literature,[6] and those readers who are interested in that methodology should consult the *International Journal of Forecasting* article cited in the footnote. This chapter focuses on the results of those audits, and the insights that we have gained about what constitutes world-class practice in demand forecasting and demand/supply integration.

Initially, the focus of the Benchmark Studies was forecasting. However, because the purpose of the ongoing research effort, as manifested in the audits, was to keep our team up to date on best practice in the field, we began to change our perspective. Over time, it became clear that a good forecast, without a good process to use that forecast to make good business decisions, was quite useless. As our team worked with the dozens of companies that participated in the research, we came to realize that our thinking needed to expand beyond documenting world-class practice in forecasting, and encompass world-class practice in demand/supply integration as well. The

[5] As of this writing (September 2012), the following 42 companies have participated in the audit research, for whom 43 audits have been completed: Eastman Chemical Corporation, DuPont of Canada, Hershey Foods USA, Michelin, Allied Signal Automotive, Exxon, Union Pacific Railroad, Lucent Technologies, ConAgra, Smith & Nephew, Ethicon, Avery Denison, Corning, Pharmavite, Motorola PCS, Williamson-Dickie Manufacturing Co., Sara Lee (Intimate Apparel Division), John Deere, Continental Tire, AET Films, Whirlpool, Michelin (re-audit), Philips North America, Bacardi, Orbit Irrigation Products, Amway, Maxtor, OfficeMax, Lockheed-Martin, Nissan, Peerless Pump, Estee Lauder, Johnson & Johnson (Vistakon division), Radio Systems Corporation, Cooper Tire, Cummins Filtration, Cintas, Tyco Electronics (Wireless Network Solutions), Wal-Mart, Winn Dixie, Mohawk Industries, Boise, Walgreens.
[6] Moon, Mark A., John T. Mentzer, and Carlo D. Smith (2003), "Conducting a Sales Forecasting Audit," *International Journal of Forecasting*, 19 (No. 1), 5–25.

result of this nearly 40 years of research, then, is a constantly evolving vision of what constitutes world-class practice in *both* demand forecasting *and* demand/supply integration. This chapter articulates that vision.

The sections that follow use the framework first articulated in the Mentzer, Bienstock, and Kahn *Business Horizons* article, and which was then expanded upon in the Moon, Mentzer, and Smith *International Journal of Forecasting* article, cited previously. In this framework, four dimensions of forecasting practice are described: functional integration, approach, systems, and performance measurement. Along each of these four dimensions, four "stages of sophistication" are articulated, ranging from Stage 1 to Stage 4. Stage 4 represents our vision of what constitutes being "World Class." In the discussion that follows, the reader is encouraged to "grade" your own organization, and see whether you can recognize which stage of sophistication you fall into. This framework was designed to be used as a diagnostic tool, and should help you to determine which area is in most need of attention. Get out your highlighter, and make note of where your company stands relative to our vision of being World Class. Don't be discouraged if you don't find yourself highlighting many Stage 4 characteristics. Of the 43 audits our research team has conducted over the past 16 years, we have yet to find a company that has Stage 4 characteristics in all four dimensions. *Every company* has opportunities for growth in some area of forecasting management. This framework is designed to help you see where your company should assign priorities in your journey toward becoming World Class.

This chapter also describes how the four dimensions of forecasting relate to the three necessary elements for DSI excellence, described in Chapter 1, "Demand/Supply Integration": *culture, process*, and *tools*. As I show, functional integration is all about *culture*; approach is all about *process*, systems is all about *tools*, and performance measurement is the use of *tools* to measure *processes*, which influences *culture*.

Functional Integration

The first dimension of demand forecasting is *functional integration*. This dimension gets at the question of "How successful is a company at getting information from those who have it to those who need it, in support of the forecasting and demand/supply integration efforts of the firm?" As noted earlier, functional integration is the dimension that truly describes the culture of the firm, specifically, the extent to which transparency, collaboration, and commitment exists for organization-wide goals. Table 7-1 summarizes the functional integration dimension, and the sections that follow describe the five themes found in it in detail.

DSI Processes

The first theme in functional integration is the extent to which the organization has successfully designed and implemented a *DSI process*. I do not repeat all the points that were discussed in Chapter 1 here, but rather encourage the reader to go back and make sure to note the important points from it. Briefly, however, I do make some observations about what constitutes a Stage 1, 2, 3, or 4 level company in terms of DSI processes. Stage 1 companies have no formal DSI process in place. In these companies, no formal forum is established that allows the demand and supply sides of the enterprise to meet and discuss their common issues and constraints. Often in Stage 1 companies, a forecast is prepared, "tossed over the transom" to the supply chain users, and no discussion takes place. In Stage 2 companies, a formal DSI process has often been defined, but execution of that defined process has either not yet reached maturity, or has "fizzled out" due to lack of appropriate change management. The most obvious deficiency in Stage 2 companies, in regard to DSI processes, is the lack of active engagement from one or more of the key players.

Table 7-1 Functional Integration

Theme	Stage 1	Stage 2	Stage 3	Stage 4
DSI processes	No formal DSI process in place.	Formal process in place, but meetings attended sporadically without decision makers in attendance.	Formal process in place, meetings consistently attended by decision makers, strong executive support.	Strong internal DSI process as well as external collaboration with key customers and large suppliers.
Organization	Each function does its own forecast.	One function owns the forecast and dictates it to other functions.	Forecasting/ DSI housed in an independent organization.	Recognized forecasting/ DSI champion in place.
Accountability	No accountability for forecasting performance.	Accountability for accuracy for forecasting personnel only.	All involved in the DSI process accountable for forecast accuracy.	Cross-functional metrics in place to drive forecasting excellence.
Role of forecasting versus planning versus goal setting	No relationship between forecasts, business plans, or goals.	Recognition that forecasts, plans, and goals are related, but plans are often made based on goals rather than on forecasts.	Forecasting drives the business planning and goal setting processes.	Forecasting, planning, and goal setting performed in an iterative fashion.
Training	Neither education nor training provided on forecasting or business planning.	Limited process and technique training provided to forecasting personnel.	Both process and technique training and education provided to forecasting personnel.	Education and training for everyone involved in the DSI process.

For example, one retail company with whom I have worked has been successful in convincing the merchandising team to participate in the DSI process, but no representatives from store operations are ever present at any of the DSI meetings. In other cases, executives allow other obligations to take precedence over the DSI meetings, and send their lieutenants to represent them. However, these lieutenants are often not able to make important decisions about either demand shaping or supply flexing, and the DSI meetings become ineffective.

In Stage 3 companies, the organization has defined and implemented a formal DSI process, and individuals who are able to make both strategic and tactical decisions consistently attend the meetings. In addition, Stage 3 companies have strong executive support for the process. At one company that participated in the audit research, the CEO made it clear that his senior leadership team was absolutely expected to attend the executive DSI meetings, with only life-threatening illness as a valid excuse to miss them. His executive team members were to "build their calendars around the scheduled executive DSI meetings." He took attendance, and sent nasty notes to anyone who failed to attend. This level of executive support worked its way downhill in the organization, and the entire series of DSI meetings were religiously attended by those who needed to be present.

A further example can help to illustrate this concept of "strong executive support." One company in our audit database was in the early stages of implementing a DSI process. However, at one of the early executive DSI meetings, the chief marketing officer of the firm boldly stated, "When sales goals and supply chain goals are in conflict, sales will always take priority." My reaction to this statement was, "They have a lot of work to do to change the culture." This statement represents siloed thinking at its best. In a DSI culture, senior executives have committed themselves to the notion of cross-functional goals. Rather than "sales goals taking priority," in a DSI culture, "profitability goals" or "shareholder value" goals are more salient, even for demand-side executives. This company had clearly not evolved to a

Stage 3 company, because strong executive support did not exist for the ideals behind DSI.

Finally, in Stage 4 companies—representing world-class characteristics—not only is there a strong *internal* DSI process in place but there is also active collaboration with important external constituents, such as large customers and major suppliers. Stage 4 companies demonstrate a culture of collaboration, both internally and externally, and information flows freely from people who have it to people who need it.

Organization

The next functional integration theme is *organization*, and by this, I mean the organizational structure that is in place to support the DSI and forecasting processes. In a Stage 1 company, each function does its own forecast, and no effective functional entity is in place to develop and distribute forecasts. We observed an extreme example of this at one company that participated in the audit research. Although the company had a forecasting group, and they did utilize a very old, outdated legacy forecasting system, the forecasts created by this group, using this system, were virtually ignored by downstream users. The situation got so bad that the logistics department in the company decided to go out and buy its own PC-based forecasting system, and create its own forecast. The procurement department got wind of this, and followed suit by buying its own PC base forecasting system (a different system). One individual in the company described it to me as "black-market forecasting." You can imagine what transpired. Each of the three forecasting systems—the "official" legacy system, and the two PC-based systems—developed different forecasts. No one was aligned, and the company was in chaos. Although this is a rather extreme example of this Stage 1 characteristic, it illustrates the problem of having no central organization of the process.

At Stage 2 companies, one organizational entity "owns" the forecasting process. This group creates the forecast—usually without effective input from other functions—then dictates that forecast to other entities without the benefit of DSI dialogue. This Stage 2 characteristic raises the question that is often asked of me by companies: "Who should own the forecasting process?" In most companies, forecasting reports to one of three different functions: sales (or marketing), supply chain, or finance. When forecasting reports to any of these three functions, structural bias can creep into the forecasting process. That bias normally takes the one of the following forms:

- If forecasting reports to sales or marketing, normally an upward bias exists. Sales or marketing executives are typically measured and rewarded based on top-line, or revenue, performance. If a forecast is biased high, then less risk exists that product won't be available when a customer is ready to buy it, and this bias would support a revenue focused culture.

- If forecasting reports to supply chain, normally a downward bias exists. Supply chain executives are typically measured and rewarded based on cost control and inventory management. If a forecast is biased low, then less risk exists that inventory will be a problem.

- If forecasting reports to finance (which is, by the way, the worst organizational structure), then the forecasts often become plan-driven forecasts. Recall from Chapter 1 that a plan-driven forecast is one where the forecast is aligned to the financial goals of the firm, whether market demand exists to support those goals or not.

The answer, then, to the question of "Where should forecasting reside organizationally?" is found in the description of a Stage 3 company. An ideal organizational structure is one where the forecasting group reports to the COO of the company and is not aligned with any particular demand, supply, or finance function. In an ideal world, each company would have a "CFO"— Chief Forecasting Officer! This individual would lead a group that is not organizationally aligned with any function, and thus not subject to the structural biases described previously.

Although theoretically, this is the ideal organizational structure, I recognize that it is not likely in practice. Forecasting has to report somewhere, and to give a more practical solution, I cite two points:

- If forecasting has to report somewhere, I recommend that it report to the demand side of the firm. In manufacturing companies, that is sales or marketing. In retail companies, that is merchandising. I come down on the side of reporting to the demand side of the company because of what forecasting is trying to do—predict future *demand*. Sales, marketing, or merchandising should have the best view of demand, and thus, forecasting should report there.
- Even if the forecasting function is not independent on an organization chart, it can be independent *culturally*. A great example of this is at a company with which I've worked for several years. This company has a forecasting organization, headed by the same individual for more than 15 years, which is perceived throughout the company as independent and unbiased. Even though this group gets moved every 2 to 3 years—from marketing, to supply chain, to strategic planning—it remains intact and *culturally* independent. This forecasting group is also responsible for managing the company's DSI process. It is an excellent example of a Stage 3 characteristic: that forecasting and DSI be housed in an independent organization.

Finally, Stage 4 companies are characterized by the presence of what we have called a "forecasting/DSI champion."[7] This individual

---

[7] Mentzer, John T., Mark A. Moon, John L. Kent, and Carlo D. Smith (1997), "The Need for a Forecasting Champion, *Journal of Business Forecasting*, 16 (Fall), 3–8.

has the responsibility for not only overseeing the forecasting and DSI processes at the firm, but also acting as an advocate for developing the culture, processes, and tools that are needed to make DSI successful. Our experience with the dozens of companies that have participated in our research has demonstrated, time and again, that without this dedicated, full-time, effective champion for forecasting and DSI, the firm has little chance of pursuing continuous improvement and achieving world-class status. When working with a company, I have a little trick that I use to see whether or not a forecasting/DSI champion is in place. During each interview that I conduct, I ask the interviewee the following question: "If your CEO wakes up in the middle of the night with a forecasting or DSI nightmare, who will he or she call?" At some companies, I will get multiple answers to this question, ranging from "He or she wouldn't have any idea who to call" to "He or she would call the VP of Sales." At other companies, I get the same answer from everyone, and the answer I get is an individual who has forecasting and DSI as his or her sole job responsibility. When this latter case exists, then I have evidence that this company is at Stage 4, and a true forecasting/DSI champion is in place.

Accountability

The next functional integration theme is *accountability*. In Stage 1 companies, no one who participates in the forecasting process is accountable for their forecasting performance. This lack of accountability is the primary driver behind some of the "game playing" that was described in Chapter 4, "Qualitative Forecasting." In Chapter 6, "Performance Measurement," I introduced the mantra of "what gets measured gets rewarded, and what gets rewarded gets done." This management proverb is clearly relevant here. When no accountability exists for the accuracy or usefulness of a forecast, then those who are responsible for providing that forecast will either spend very little time and energy to do it well, or will use the forecasting process to advance the agendas that I described in Chapter 4. In terms of accountability,

I emphasize that Stage 1 companies fail to impose accountability for forecasting *performance*. At one company with which I worked, the individuals from the sales team were paid bonuses if they completed their forecasts *on time*. It didn't matter whether their forecasts were accurate—they just had to be in on time. That's still a Stage 1 company.

Stage 2 companies introduce accountability for forecasting performance, but that accountability is limited to those who work in the forecasting group. In these companies, accuracy is measured and the forecasters have goals they strive to achieve. There are rewards for achieving those accuracy goals, and consequences for not achieving them. While this is better than Stage 1, it is suboptimal because it does not extend the accountability to others in the company, such as sales, marketing, or merchandising, who are involved in the forecasting process. This extended accountability is present in Stage 3 companies. Here, each individual who is involved in the process has his or her contribution to forecast accuracy measured. Again, "What gets measured gets rewarded, and what gets rewarded gets done." If a company wants, for example, its salespeople to do a good job of forecasting, then their contribution to the forecast must be measured, and they must be accountable, through rewards and consequences, for good or bad performance.

Finally, Stage 4 companies recognize that forecast accuracy is a process measure, not an outcome measure. These companies work to affect the overall culture of the firm by incentivizing their employees, particularly their senior functional leaders, with cross-functional metrics. For example, at one company with which I've worked, the sales organization has begun including finished good inventory levels in the overall performance metric of their sales executives. Finished good inventory is an outcome metric. This individual, then, must create a culture where his sales team provides accurate, credible forecasts, a process metric, so that the firm can make good cross-functional decisions that keep finished goods inventories at healthy levels. This is an example of a Stage 4 company, applying cross-functional performance reward to those involved in the forecasting process.

Role of Forecasting versus Planning

The DSI process described in Chapter 1 is, at its essence, a planning process. The point was made at that time that DSI is most effective when it is positioned as the way to "plan the business," rather than "plan the supply chain." The vision of world-class demand/supply integration articulated in Chapter 1 is one where *forecasts serve as inputs to business plans.* In other words, a forecast—the best guess about what future demand will be, as well as the best guess about future supply capabilities—leads to a plan, which is a set of decisions about what to actually *do.* Further, every organization establishes many business goals—targets that individuals and organizations strive to achieve, and for which rewards are frequently given for successful attainment and consequences suffered from non-attainment. Again, following the vision of world-class demand/supply integration, goal setting is at least influenced by a realistic assessment of the true demand in the marketplace.

In Stage 1 companies, these three concepts are at best poorly coordinated, and at worst, unrelated. In these companies, plans are often formulated based only on the financial goals of the firm, rather than a dispassionate analysis of true demand in the marketplace. As discussed in Chapter 1, this is referred to as "plan-driven forecasting," and it is one of the most insidious aberrations to an ideal DSI process. A good example of this phenomenon occurred at one company that participated in the audit research. At this company, the demand planners described a situation where they would work diligently all month to complete what they thought was a supportable, evidence-based forecast. However, at the consensus forecasting meeting, the senior executives would look at this forecast and say, "No, that's not enough. We won't make our numbers with this forecast. Raise everything by 10%." In frustration, the forecaster commented to me, "Why don't they just tell us what number they want at the beginning of the month? Then, I could spend my time doing something useful, like playing golf!" At this company, and at several other companies with which we've worked, either no relationship, or a dysfunctional relationship, exists between demand forecasting, business planning, and goal setting.

Stage 2 companies are characterized as having processes in place that attempt to appropriately align forecasting with business planning, while separating forecasting from goal setting. However, in spite of these processes, the business plans often end up being formulated based on the firm's goals, rather than forecasted demand in the marketplace. In other words, it's almost as if the company knows that it shouldn't, but when the end of the fiscal quarter or the fiscal year looms, the company almost can't help itself from forecasting the goal, then hoping for the best.

In contrast to these problematic cultural inclinations, Stage 3 companies have processes in place that are rigorously followed to formulate business plans following careful consideration of the demand forecast. And then, after a company reaches Stage 4, a truly iterative process is followed that is consistent with the ideal state of DSI described in Chapter 1. Both demand forecasts *and* financial goals serve as input to the overall planning process. If the amount of demand in the forecast is insufficient to reach the financial goal, then various "gap closing" strategies are explored, and the most financially sound, and strategically aligned, gap-closing strategy is selected during the DSI process. This strategy is then translated in specific action steps, which ultimately constitute the business plan. In addition, goals are set by combining the dispassionate assessment of future demand found in the forecast with the strategic growth objectives of the firm into a set of targets that contributes to the firm's overall motivational efforts.

Training

Functional integration is that dimension of forecasting and DSI that relates most closely to *culture*. Many elements can influence culture, including organizational structure and accountability, discussed earlier. One other important way to effectively drive a culture of openness and collaboration is through extensive education and training of the individuals involved in the DSI process. I use both terms—education and training—on purpose, and don't see them as synonymous. In simple terms, I use *training* to describe efforts to teach people how to *do things*. In contrast, I use *education* to describe efforts to teach people how to *think about things*. Both are important, but to change culture, education might be more important than training.

At Stage 1 companies, neither education nor training is effectively utilized in the forecasting and business planning processes. In these companies, forecasters are frequently only shown how to operate the forecasting system, and which sequence of steps to complete to end up with a demand forecast in advance of a certain deadline. The only semblance of training is directed at the forecasting personnel—no one else in the firm who contributes to forecasts, such as salespeople, product managers, or marketing people, receive any training or education whatsoever in these business processes. The result of this lack of training or education is frequently that people fail to understand the purpose of forecasting, how a forecast should be created, or what happens to the forecast after it is completed. Moving an organizational culture toward openness and collaboration is nearly impossible when key individuals know little about the how or why of forecasting and demand planning.

Stage 2 companies are those that strive to improve the forecasting process by focusing on training—focusing on the "how"—for their forecasting personnel. At these companies, training on the mechanics of statistical forecasting is provided, along with the steps needed to acquire and incorporate qualitative judgment into the forecasts.

Because the training delivered at these Stage 2 companies tends to be quite tactical, often limited time is spent helping the forecasting personnel understand the strengths or weaknesses of various statistical forecasting techniques. Although Stage 2 companies do provide some enhanced opportunity to drive culture through training, that opportunity is limited because it is centered on the "how" rather than the "why," and it is delivered to the forecasting personnel only. To quote an old cliché, it is to some degree "preaching to the choir."

Stage 3 companies are more committed to extensive education and training for forecasting personnel. At these companies, much more time is spent educating forecasting personnel on the ideas behind various forecasting tools and techniques, both quantitative and qualitative, and to help these individuals know what pitfalls to expect from different approaches. At one company in our audit database, significant investment has been made for many years on providing formal certification to its very large forecasting team, located worldwide. Videotaped lectures are combined with formal exercises and reading material, and the curriculum culminates with an examination designed to demonstrate mastery of the material. Both the "how" and the "why" are emphasized in this certification process, and by the time its demand planners have concluded the program, they are recognized experts. Although this company excels at delivering education and training to its demand planners, they are not world class. In order for this, or any company, to reach Stage 4 and be truly world class on this dimension, extensive education and training must be provided to *everyone* who participates in the DSI process. In other words, not only forecasters need education and training. If salespeople, product managers, marketing managers, even senior executives, participate in the process, then all these individuals need to know *how* to do what the process requires them to do, as well as *why* they need to do it. When companies embrace this commitment to education and training for everyone, they take advantage of the opportunity to truly drive the type of organizational culture that needs to be in place for DSI to work well. Recall the example from earlier in this chapter of the marketing executive who demonstrated siloed thinking when he stated "If sales goals and supply chain goals are in conflict, sales goals will always take priority." A well-designed and well-executed education effort (along with changes in the measurement and reward structure!) is needed to transition this company away from its siloed culture into one where true demand/supply integration drives the company.

Summary: How Companies Can Improve on the Functional Integration Dimension

Recall that the functional integration dimension answers the question, "How successful is a company at getting information from those who have it to those who need it, in support of the forecasting and demand/supply integration efforts of the firm?" It is the dimension that is all about organizational culture. If, while reading this section, you have identified your company as being in Stage 1 or 2 on several of these themes, here are some bullet points to focus on:

- **Work on your DSI processes, and gain committed executive support.** A point that I have made repeatedly in this book is that getting demand-side executives (sales and marketing in a manufacturing environment, merchandising in a retailing environment) to go "all-in" on demand/supply integration is extremely important. Identify a demand-side executive champion, and leverage that individual's influence to execute a well-conceived change-management strategy to change the culture.

- **Pay attention to the organizational structure.** Try to establish a truly independent forecasting function, whose sole agenda is to create the most accurate, unbiased demand forecast possible to drive the business planning process. Identify a forecasting/DSI champion whose full-time job is to effectively drive process improvement and culture change.

- **Make everyone involved in the process accountable for his or her contribution.** "What gets measured gets rewarded, and what get rewarded gets done."

- **Make sure that everyone understands the difference between forecasting, business planning, and goal setting.** Use organizational structure, accountability, and education and training to ensure that these processes are sequenced properly: forecasting drives the business planning and goal-setting processes, and all should be done in an iterative fashion.

- **Commit resources to effectively train and educate everyone involved in DSI on the "how" and the "why" of what they're doing.**

One final word of caution, and exhortation, concerning functional integration: In our experience working with dozens of companies, it's the most difficult dimension to influence. Culture is very hard to change. However, it's the dimension that provides the biggest pay-off if it moves toward being world class. Getting everyone in the boat, pulling on the oars together, has more value than any other element of demand forecasting and demand/supply integration.

Approach

The second dimension of demand forecasting is *approach*. This dimension gets at the question of "What are the techniques and processes that are used to create demand forecasts?" As noted earlier, excellence comes from *culture, process*, and *tools*, and approach is the dimension that describes the *process* element of forecasting and demand/supply integration excellence. Table 7-2 summarizes the approach dimension, and the following subsections describe each of the five themes found in it.

Forecasting Point of View

The *forecasting-point-of-view* theme describes the perspective that is taken on creating a forecast. Stage 1 companies look at the forecast through the lens of the financial goals of the firm, or *plan-driven forecasting*. Several sections of this book cover the insidious nature of plan-driven forecasting and thus, I do not dwell further on it now. Suffice it to say that when a forecast is developed by looking through the lens of the financial goal of the firm, it is not good. It's Stage 1, and improvement is clearly needed.

In Stage 2 companies, the lens through which the forecasting process looks is a "bottom-up" lens. This perspective entails essentially beginning with customer #1 and asking "What will their demand be?" then moving on to customer #2 and asking the same thing, then on to customer #3, and so on, through all the customers, and then adding all these individual forecasts together to reach an overall projection. A variation of this would be to individually forecast only those large, "A"-level customers, and then group smaller customers into an "all-other" category, and use a statistical modeling technique to forecast these smaller customers. The sum of the "A" level customer forecasts and the "all-other" forecast would then constitute the demand forecast for that SKU, product family, brand, or whatever forecasting level is currently being analyzed. This straightforward process provides a valuable insight. It does, however, have potential pitfalls. One pitfall is best described through two examples from companies that have participated in the audit research. The first company was a manufacturer of optical fiber, those human-hair thickness strands of glass that form the backbone for the transmission of digital signals worldwide. As described in Chapter 2, "Demand Forecasting as a Management Process," the forecasting form for optical fiber is kilometers—in other words, the relevant question asked during the forecasting process is "How many kilometers of optical fiber will be demanded in future

Table 7-2 Approach

Theme	Stage 1	Stage 2	Stage 3	Stage 4
Forecasting point of view	Plan-driven forecasting orientation.	Bottom-up forecasting orientation.	Both top-down and bottom-up forecasting orientation.	Top-down and bottom-up forecasting, with detailed reconciliation.
What is being forecasted	Shipments are the basis for historical analysis.	Shipments plus backorder adjustments are the basis for historical analysis.	Shipments plus backorder adjustments plus self-recognized demand are the basis for historical analysis.	Complete visibility of sell-in and sell-through allows modeling of true demand.
Forecasting hierarchy	No hierarchy in place.	Partially defined hierarchy in place; one or two of the three "faces" defined.	Complete hierarchy in place; all three faces defined.	Complete hierarchy in place, with reconciliation possible between levels and across faces.
Statistical analysis	No statistical analysis of historical demand.	Limited use of simple statistical models.	Appropriate use of time-series models.	Both time-series and regression models used appropriately.
Incorporation of qualitative input	Little to no qualitative judgments used to augment statistical forecasts.	Sales, marketing, and senior executives participate in the process, but their input is often highly biased because of other agendas.	Sales, marketing, and senior executives provide unbiased judgment, leading to appropriate adjustment of statistical forecasts.	Both internal and external (for example, customers) sources of qualitative judgment are incorporated into the forecasting process.

time periods?" The direct customers for the optical fiber company are companies who are called "cablers," because their value-add in this supply chain is to purchase the optical fiber, then bundle some number of strands of fiber together, wrap protective material around these fragile strands of glass, and create optical fiber cable that can then be sold to telecommunications companies, who eventually bury that cable, thus providing digital transmission capability. The optical fiber company's salespeople thus call on cablers, and their job is to forecast demand from these cablers. One of the demand planners from the optical fiber company told me a story that illustrates a pitfall associated with bottom-up forecasting. Imagine a demand planner approaching a salesperson and asking for his or her forecast of demand for the next quarter. The salesperson might respond, "Well, my customer, Cabler A, is very confident that they will win the contract for optical fiber in South Korea, so they will need 40,000 kilometers of fiber in the next quarter." The demand planner notes that forecast, and moves on to the next salesperson. The next salesperson reports "My customer, Cabler B, is very confident that they will win that South Korea contract, so put me down for 40,000 kilometers of fiber." Now with a puzzled look on her face, the demand planner goes to the next salesperson who declares "My customer, Cabler C, is counting on winning that contract in South Korea, so I'll need 40,000 kilometers of fiber." The astute reader can instantly see the problem. Only one of these three cable companies is going to win that South Korea contract, and if the demand planner follows a strict bottom-up procedure, the fiber optical company will overforecast by 80,000 kilometers.

The second example, similar to the first, was described during an audit of a candy manufacturer's forecasting processes. Here, the candy company sells its products to retailers, who then sell the candy to consumers. In this case, the conversation between forecaster and salesperson might go something like this: "My customer, Retailer A, is really going to be pushing candy next quarter because they believe it's a very high margin item, so they expect to grow volume by 5%." The next salesperson says, "My customer, Retailer B, has declared candy to be a strategic focus for the next quarter, so their demand will increase by 5%." The next salesperson might say the same thing about Retailer C. Each retailer expects its demand for candy to grow by 5% in the next quarter. However, a close examination of overall industry demand for candy might reveal flat demand for the past several quarters, leading the forecaster to believe that if Retailer A is going to grow share, it's going to do it at the expense of Retailer B or Retailer C, and that all three will grow by 5% is unlikely. What do you do?

The solution is to do what Stage 3 companies do, which is to take both a bottom-up and a top-down perspective on the forecast. A bottom-up forecast looks customer by customer, whereas a top-down forecast requires two projections. First, the forecaster must project *overall industry demand* for the product or service category being forecasted. Second, the forecaster must project his or her company's *market share* of that industry demand. Multiply those two numbers together, and the result is a top-down forecast of demand. In both examples just described, a top-down forecast would give a very different number than the bottom-up forecast. Stage 3 companies perform both analyses, and Stage 4 companies take the analysis one step further by conducting a detailed reconciliation of the differences between the top-down and the bottom-up perspective. Such a detailed reconciliation not only helps the forecaster arrive at a more useful result, but also gives the forecaster valuable insights about business dynamics, making him or her a much more valuable contributor to the overall DSI process.

What Is Being Forecasted?

The second theme in the approach dimension concerns the issue of forecasting *true demand*. As defined in Chapter 2, true demand is "what customers would buy from us if they could." If the company has sufficient service or manufacturing capacity, no orders are ever unfilled, and all orders are supplied at the time and quantity requested by the customer, then "how much we have shipped" would be identical to "how much was demanded." However, this ideal is seldom reached. Thus, the second theme in the approach dimension describes the source of historical data against which statistical, or quantitative analysis, takes place.

Stage 1 companies use historical shipments as the basis for statistical analysis of past demand, which is then projected into the future to forecast future demand. This procedure is certainly quite easy to implement. Pulling actual shipment records, putting those shipments into monthly buckets, and analyzing those monthly buckets to find patterns that can be projected into the future is straightforward. Revisit Chapter 2 for a refresher the pitfalls of using shipments as a surrogate for actual demand. Recall the example of the chemical company forecasting sodium benzoate. In that example, the company was unable to supply the product demanded by the customer in the quantities demanded, in the time required. As illustrated there, the shipment of 15,000 pounds of sodium benzoate in August, along with the 5,000 pounds in September, did *not* reflect true customer demand. *True demand* was 20,000 pounds in August, even though the company *shipped* 15,000 in August and 5,000 in September. A Stage 2 company solves this problem fairly easily by making an adjustment to the shipment record. Here, the chemical company can easily identify that the requested ship date for all 20,000 pounds was August, and then use that number as the basis for their forecast of future demand.

But if you recall the next example in Chapter 2, where the customer decides not to take the 5,000 pounds in September because they really needed all 20,000 pounds in August, no order record exists for the full 20,000 pounds demanded. In this scenario, a Stage 2 company will do an incorrect forecast, but a Stage 3 company will be more sophisticated, and have created a "lost order" record, which is referred to in Table 7-2 as "self-recognized demand." This Stage 3 company will now have the number that recognizes true historical demand—20,000 pounds—that it can now use to run its statistical analysis, and project that pattern of demand into the future. Thus, Stage 3 companies do a better job of accessing "true historical demand" than do either Stage 1 or 2 companies.

Companies that reach Stage 4 in this theme are those companies who sell their products through channels of distribution. World-class practice in this setting is to have complete visibility of both sell-in and sell-through, which allows the modeling of potential out-of-stock situations, and allows the company to estimate true demand at the consumer level. *Sell-in* refers to the quantity of product that is demanded by the retailer, which might be placed in inventory at the retailer's distribution center, in individual stores' back rooms, or on the shelf at the retail outlet. *Sell-through* is the quantity that actually passes through the store's cash register, and this data is often provided to manufacturers by their retail partners in the form of POS (Point-of-Sale) data. If this sell-through data is available, it can be analyzed alongside the sell-in data, which is simply drawn from the manufacturer's internal demand data. When sell-in greatly exceeds sell-through, then inventory is being built up. When sell-through exceeds sell-in, then inventory is being depleted. If retailer inventory is being modeled in this way, then the manufacturer can pinpoint possible out-of-stock situations at the shelf or distribution center, and anticipate situations where retailer demand is likely to rise or fall as a part of its inventory management policy.

Forecasting Hierarchy

Chapter 2 describes the concept of the *forecasting hierarchy*. Recall that the hierarchy is a way of describing the levels of granularity at which a forecast can be expressed. For example, a product forecast can be conducted at the SKU level, which is the lowest level of granularity; at the brand level, which is one level up from SKU;

at the product category level, which is one level up from brand; or at the overall company level. These designations are idiosyncratic to each company, and the terminology will vary as well. For example, at several companies in our audit database, the lowest level of granularity is the "part number" level, and at others, a "product family" level falls between the brand and the SKU levels. However the levels are operationalized, the concept is the same—the forecasting hierarchy describes how lower level forecasts are aggregated, or "roll up," to higher-level forecasts. Recall also that the forecasting hierarchy consists of three separate "faces": the product, location, and customer face. As discussed in Chapter 2, the idea is that people should be able to contribute information to the forecast at the level of granularity in which they know it, and take information out of the forecast at the level of granularity in which they need it. Typically, the hierarchy is executed in the firm's backbone ERP system, where the "parent-child" relationships between levels of products, customers, and locations are defined in the system.

At Stage 1 companies, no defined forecasting hierarchy is in place. If one function in the company needs a forecast at the SKU level, and a different function in the company needs a forecast at the brand level, separate forecasting processes must be completed for each needed forecast. This Stage 1 characteristic is most commonly a by-product of failure to either install an Enterprise Resource Planning (ERP) system, or failure to bring the organization's ERP system up to date. In contrast, Stage 2 companies have at least one face of the hierarchy in place. Usually, if a company has only one of the three faces defined in its system, it will be the product face that is present. However, the customer and location faces are equally necessary to adequately plan for both distribution and sales coverage. In Stage 3 companies, all three faces of the forecasting hierarchy are operational in the system, and Stage 4 consists of a fully operational three-face hierarchy, in which data easily "flows" between the levels and across the faces. In other words, a company that is at Stage 4 in this area is truly able to "put information in at the level that people know it," as well as to "take information out at the level that people need it."

Statistical Analysis

An entire chapter in this book, Chapter 3, "Quantitative Forecasting Techniques," covers the topic of quantitative, or statistical, forecasting. Unfortunately, in practice some companies operate at Stage 1 by not taking advantage of the insights that are gained from examining historical demand, and using statistical tools to find patterns in that history. Of course, in some instances statistical forecasting does not provide much benefit. (Remember the example of Boeing forecasting demand for 747s?) However, a company should see itself as Stage 1 where in fact insights can be gained from this analysis, but the company does not perform it. Our team was shocked when, at one of the very first companies we audited, we discovered that even though it sold products that had clear historical trends and seasonality, the only approach that was used to forecast was to send out forms to the salespeople, which were essentially blank pieces of paper, asking the question, "How much do you expect to sell over the next quarter?" Clearly, this company was mired in Stage 1!

Stage 2 companies make an effort to use some statistical tools to forecast demand, but the tools used are fairly elementary, and often used inappropriately. For example, if a company only calculates the average number of units demanded per month over the last year, and uses that one number as its forecast for the next year, that would be consistent with a Stage 2 company (unless, of course, the historical demand pattern consists only of random noise, in which case, an average is as useful a technique as can be found). Another example would be for a company to use a 6-month moving average where seasonality repeats every four months. As described in Chapter 3, this would simply be an inappropriate choice for a statistical model to choose and would constitute Stage 2.

After a company has progressed to Stage 3, it is appropriately using the full range of time-series tools to most accurately discover patterns in historical demand, and then project those patterns into the future. Typically, when a company reaches Stage 3, it is using sophisticated forecasting software that includes an array of time-series algorithms, and allows "pick-best" functionality where the system chooses the "best" algorithm to match the historical demand. Of course, a caution always applies to the use of pick-best functionality, as illustrated in Chapter 3 in the "St. John's Wort" example. Those companies that have reached Stage 4—world class—are using not only all the time-series techniques at their disposal, but also using regression or other causal modeling tools to uncover the relationships between independent variables and demand. In our experience, the companies that benefit most from this level of statistical analysis are those companies that are very promotion-intensive. These firms gain tremendous benefit from documenting the "lift" that occurs from different promotional strategies, and then projecting that lift into the future when similar promotional activities are scheduled. These firms also benefit from the strategic insights that can be derived from analyzing which promotional activities have historically generated the most lift under specific conditions.

Incorporation of Qualitative Input

As discussed in Chapters 2, 3, and 4, the sequence of steps that under most circumstances results in the most satisfactory demand forecasting outcomes is one where a statistical forecast is augmented by qualitative judgments from knowledgeable people. In other words, *the first step* should be a statistical analysis of historical demand to try and understand what patterns have existed in the past, then projecting those patterns into the future. *The next step* is for knowledgeable people to examine those statistical projections and make qualitative adjustments based on their judgment of how the future is likely to look different from the past.

Stage 1 companies make little to no effort to augment statistical forecasts with qualitative judgments from sales, marketing, or senior executives. We observed a good example of this situation in one company in our audit database, where soon before our engagement with the company, significant expenditures had been made on a statistical forecasting system. An unanticipated consequence of this installation was that the sales and marketing teams stopped contributing to the forecasting process. As we learned when we interviewed sales and marketing people, their perspective had become "Look, we spent all this money on the forecasting system, now I don't need to be involved. Let the system do the forecast!" Overall forecasting performance, as measured by MAPE, plummeted, and the company blamed the new system. In fact, it was the overall process that had deteriorated.

Stage 2 companies are characterized by active participation, from sales, marketing, and senior executives, but that participation sometimes does more harm than good because of various other political agendas. As discussed in detail in Chapter 4, it is often the case that sales, marketing, product management, or senior executives—the most common contributors of qualitative forecasts—will have goals other than forecast accuracy that drive their efforts. I won't repeat that discussion here, but instead encourage the reader to revisit that section of Chapter 4. Rigorous measurement of forecasting outcomes, and reward structures that recognize the importance of accurate forecasts, are needed to move a company beyond Stage 2, and into Stage 3. Stage 3 is characterized by both active and helpful participation from sales, marketing, product management, and senior executives in the forecasting process. In these companies, processes are established to help these individuals contribute to the forecast in a way that takes advantage of what they know well, while not making their participation onerous. Normally, this occurs through the demand review in a well-structured DSI process, as described in Chapter 1.

When companies reach Stage 4 in this area, they are effectively utilizing not only qualitative insights from internal sources, but from external sources as well. The most common external source of qualitative judgment is the customer base. As discussed in Chapter 5, "Incorporating Market Intelligence into the Forecast," companies can gain considerable insight from collaboration with those customers who are both willing and able to provide that level of insight. Although insights about customer demand can come through the sales teams, having forecasters work directly with customers to better understand potentially changing demand patterns often provides value. This is particularly helpful in a retailing context, where a manufacturer is selling its products through retail customers. Direct collaboration between the procurement organization at the customer and the demand planners can often be an extremely effective way to get that customer insight into the demand forecast. As discussed in Chapter 5, these collaboration processes can be fairly informal, or very formal as is the case in a CPFR relationship. However they are structured, close collaboration with key customers in capturing their anticipated future demand takes a company to world-class level in the approach dimension.

Summary: How Companies Can Improve on the Approach Dimension

To review, the approach dimension answers the question "What are the techniques and processes that are used to create demand forecasts?" If functional integration is all about *culture*, then approach is all about *process*. If, while reading this section, you have identified your company as being in Stage 1 or 2 on several of these themes, here are some bullet points to focus on:

- **Both top-down (industry level) and bottom-up (customer-by-customer) forecasting is useful.** Your company should do both. You should also work to understand the gaps between the bottom-up and top-down views (and usually gaps exist), because this can both help you do a better job of forecasting, and help you understand the dynamics of your business better.

- **Do everything possible to forecast *demand*, not *sales*.** Knowing what your historical demand actually was is not always easy, but finding the best surrogate possible for true demand is worthwhile.

- **Create, and use, a forecasting hierarchy.** If you are using Excel or some other desktop spreadsheet tool as your main forecasting engine, then creating and using an effective forecasting hierarchy will be very difficult. You need the power of well-conceived and executed data structures to work effectively with various levels of the forecasting hierarchy. The goal is to be able to "put information in at the level you know it, and take information out at the level you need it."

- **Take advantage of the power of statistical forecasting.** The past is an excellent starting point for your forecast of what you think will happen in the future. Companies can gain much from the appropriate use of both time-series and regression or other causal modeling tools.

- **Combine the power of statistical forecasting with the wisdom that comes from the judgment of individuals.** Create incentives for people to contribute useful judgments that will help make the overall forecast more accurate, and less biased.

Although the approach dimension might not be as difficult to improve as functional integration, because it is less dependent on culture, it is difficult enough. Frequently, effectively changing the approach dimension is hampered by the "That's the way we've always done it" syndrome. Inertia can be a powerful enemy, and getting people to change the way they do things can be tough. At the same time, when combined with the culture changes that come about from improving functional integration, the process improvements companies gain from moving to world-class level in the approach dimension can greatly enhance demand forecasting, DSI, and overall business performance.

Systems

The third dimension of demand forecasting is *systems*. This dimension gets at the question of "What is the state of information technology support for the demand planning process?" Recall that excellence comes from *culture, process,* and *tools,* and systems is the dimension that describes the *tools* element of forecasting and demand/supply integration excellence. Table 7-3 summarizes the systems dimension, and the following sections describe each of its four themes in detail.

In Chapter 2, I rather vehemently stated, "Systems are not silver bullets." My point in making that statement was that a company cannot "buy its way" into forecasting excellence by installing a forecasting system. Elsewhere, I've stated that culture represents 60% of the challenge in creating excellence in forecasting and demand planning, process represents 30%, and tools only 10%. Although tools—systems—only represent 10%, they are important and I don't want to suggest otherwise.

To explicate some of the issues in the systems dimension, see Figure 7-1, which is a figure that was previously discussed in Chapter 2.

Table 7-3 Systems

Theme	Stage 1	Stage 2	Stage 3	Stage 4
Level of integration	Separate systems require manual transfers of data.	Separate systems are linked by customized interfaces.	Partially integrated supply-chain suites are in place.	Integrated software suites allow external data sharing with suppliers and customers.
Access to performance measurement reports	No performance metrics are calculated anywhere in the forecasting system.	Performance metrics are calculated in separate systems.	Performance metrics are calculated and available online.	User-friendly report generator allows for performance reports on demand.
Data integrity	"Islands of analysis" are prevalent.	Inconsistent databases distributed over multiple systems require frequent data manipulation.	Demand history resides in a central database, updated in a batch process.	Data warehouse design supports demand history updates in real time.
System infra-structure	Investments in hardware, software, and IT support are inadequate.	There is minimal system infrastructure support for hardware, system maintenance, and upgrades.	Acceptable system infrastructure is provided for system maintenance and upgrades.	Superior system infrastructure permits timely upgrades and enhancements needed to support improved forecasting management.

Level of Integration

Figure 7-1 shows a simple overview of an organizational system structure in which demand forecasting systems are tied directly to the firm's ERP backbone system, which is then tied directly to the firm's data warehouse. In Stage 1 companies, the level of integration that is implied by Figure 7-1 is not present. Rather, historical demand data that is needed by the forecasting system is not easily accessible from a data warehouse. Instead, data must be transferred manually from the central data repository to the forecasting system. Similarly, at Stage 1 companies, after the forecast is complete, the results must be transferred manually into the systems that use the forecast for planning purposes, such as Materials Resource Planning (MRP) systems, production planning systems, or inventory management systems. Sometimes, data is transferred manually, with numbers literally being keyed in by data entry personnel. As someone who at one time worked in the data processing industry, visiting a company and seeing a computer printout sitting next to someone's workstation, with that individual keying numbers that were extracted from one system into another system, is painful. More commonly, however, is a situation where an analyst will "cut" data from one spreadsheet and "paste" that data into another spreadsheet. For example, a SKU-level forecast might be completed, sometimes in a spreadsheet program, and the results are cut-and-pasted into the production planning application, which might be a different spreadsheet. Although certainly faster than data rekeying, a cut-and-paste procedure can potentially lead to big mistakes. Cutting a 100 by 100 block of data in one spreadsheet and pasting it into the wrong place on another spreadsheet results in 10,000 mistakes from one bad keystroke! The results of being Stage 1 in this dimension should be quite clear. Manual transfers of data, either through cut-and-paste or rekeying, is highly error prone and a huge waste of time.

In Stage 2 companies, the manual transfer of data is replaced by customized interfaces that link separate systems together. In these situations, there might be a variety of downstream supply chain systems that use the output of the forecasting system, but the data transfer between these systems takes place through custom code. Several problems are associated with this approach. One is that completing these custom interfaces takes considerable time and resources. Also,

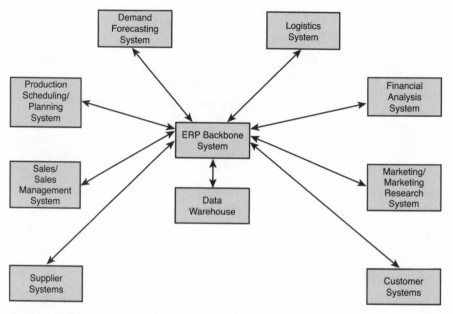

Figure 7-1 Forecasting system overview

after those custom interfaces are completed, companies become reluctant to upgrade the systems on either end of the interface, because an upgrade might also require rewriting the customized interface code. Finally, over time, these interfaces often become quite complex. "Cludgey" is the word that was used to describe such interfaces back in my days in the computer industry (the opposite of cludgey is "elegant"). Such complex interfaces are often quite slow, reducing the computing power of the forecasting system. I have even observed instances where the interfaces are so complex that only one person really understands how they work. If that one person decides to retire, gets hit by a bus, or gets hired away by a competitor, then the company is in deep trouble, because no one else really understands how the data move into, or out of, the forecasting system.

Stage 3 companies are characterized by elegant, rather than "cludgey" interfaces between the various forecasting and supply chain systems. Often, a single software vendor provides the functionality contained in these systems, and the elegant interfaces are engineered into the integrated system. Also, some forecasting system vendors engineer elegant interfaces both into, and out of, the backbone ERP systems, knowing that customers are unlikely to choose their products unless those interfaces are solid. This level of integration is taken to an even higher level at Stage 4 companies, where not only does tight integration exist between the internal forecasting and supply chain systems, but also integration between internal system and external customer and supplier systems. Our discussion of external DSI in Chapter 1 showed information flows between various members of the firm's supply chain. An OEM customer's demand plan becomes input to a component manufacturer's demand forecast. A component manufacturer's operational plan becomes input to the OEM customer's capacity forecast. At Stage 4 companies, this information flow is seamless.

Access to Performance Measurement Reports

Chapter 6 discussed why measuring forecasting performance is so important. Forecast accuracy is a common surrogate measure for demand volatility, which is an important variable in the calculation of safety stock inventory. Accuracy metrics help to validate the usefulness of the forecasting processes and techniques that have been chosen by the demand planners. Measuring accuracy, and tracking it over time, provides critical information for managers to assess overall process improvement efforts. Plus, without good accuracy metrics, providing appropriate rewards for forecasting excellence, or consequences for forecasting incompetence becomes difficult for managers. However, none of these benefits are available unless access to performance metrics exists somewhere in the forecasting system. In this theme, the difference between "best in class" and "worst in class" is tied to the accessibility of forecasting performance data by people who need those metrics to drive excellence.

Stage 1 companies—worst in class—do not have performance metrics calculated anywhere in the forecasting system. That is not to say that these metrics are not calculated anywhere in the company. At some companies that we have classified as Stage 1, individual users calculate their own metrics for their own personal use. However, no effort exists to centrally calculate and distribute these performance metrics to people who need them. Without this central control over performance metric calculation, comparing performance across products, or customers, or regions is impossible. Stage 2 companies, in contrast, do make performance metrics available to those who need them, but they do so in such a way that makes the distribution of these reports cumbersome. Typically, Stage 2 companies use tools such as Access and Excel to extract data from the forecasting and ERP systems, dump those data into a spreadsheet, and then calculate performance metrics. These procedures are normally performed offline, and the resulting performance metrics are only available in printed reports or on spreadsheets that are not very customizable, and thus, not very user friendly.

More sophisticated companies that we would classify as Stage 3 use the forecasting systems themselves to calculate performance metrics. Most respectable forecasting systems contain the ability to calculate these metrics directly in the system, and then, when appropriate access privileges are given to designated users, those users can access those metrics directly from the forecasting system. World-class companies—Stage 4—make use of highly user-friendly report generators that allow designated users to craft their own performance metrics to suit their own needs. For example, at Stage 4 companies, sales managers can query the system to calculate an individual salesperson's forecast accuracy for a particular product family at a particular customer, and compare that accuracy to the average of all salespeople in the company. An inventory manager could query the system to receive a 12-month rolling MAPE for a particular SKU at a particular location, which he or she could then use to calculate the appropriate safety stock levels for that SKU at a distribution center. In other words, Stage 4 companies give access to performance metrics to people in the format they need it, when they need it.

Data Integrity

Mentzer and Kahn, in their 1995 *Journal of Forecasting* article,[8] reported that the most important outcome of a forecasting process for forecast users was accuracy. Not far behind, however, was credibility. In other words, forecast users want the forecasts given to them to be accurate, but they also want to believe in the integrity of those forecasts. One of the biggest threats to credibility is a lack of data integrity. Stage 1 companies struggle with data integrity because they find themselves in an "islands of analysis" situation. "Islands of analysis" is a term coined by Mentzer and Moon[9] to describe a situation where multiple people are using different, unconnected, and uncoordinated tools to forecast demand for different products or customers. "Islands of analysis" is often a problem when a company uses Excel or some other spreadsheet tool that resides on individual analysts' desktops to forecast demand. At one company with which we worked, the islands of analysis problem was so prevalent that one observer described his company as suffering from "spreadsheet mania." Several reasons exist for why having islands of analysis is suboptimal. For one thing, over-reliance on spreadsheets makes centrally managing a forecasting hierarchy difficult. For another, little central control exists over the techniques used to model historical demand. Perhaps the most daunting problem posed by islands of analysis, or spreadsheet mania, is a lack of data integrity. As discussed in Chapter 2, and illustrated in Figure 7-1, a well-conceived system architecture is one that utilizes

---

[8] Mentzer, John T. and Kenneth B. Kahn (1995), "Forecasting Technique Familiarity, Satisfaction, Usage, and Application," *Journal of Forecasting*, 14 (No. 5), 465–476.

[9] Mentzer, John T. and Mark A. Moon (2004), *Sales Forecasting Management: A Demand Management Approach*, Thousand Oaks, CA: Sage Publications.

a professionally managed data warehouse to control access to, and integrity of, the company's data. After that data is placed on individual analysts' desktops and manipulated by individual analysts' spreadsheets, data integrity can become compromised, which leads to a lack of credibility in the forecasts.

At Stage 2 companies, "islands of analysis" might not be a problem, but data integrity can be problematic if a variety of databases exist where historical demand data resides, and where data manipulation is required to access this historical data on multiple systems. One instance where our research team has seen considerable problems in this area is at a company that has aggressively grown through acquisitions of other companies. At this company, multiple ERP systems are in place supporting multiple data warehouses. Significant translation of data must take place on a regular basis, primarily so that part numbers in one system match up with part numbers in another system. This type of repeated data manipulation leads to delays, errors, and ultimately loss of confidence in the integrity of the forecasts.

In Stage 3 companies, the type of architecture illustrated in Figure 7-1, which shows a central, professionally managed data warehouse, is in place. A common practice at these companies is for this data warehouse to be updated in a batch process, often once per day, with new forecast or actual demand information. In other words, orders are placed and processed throughout the day, but once a day, the central data warehouse is updated with the new information. In most situations, this type of updating is adequate. However, in some settings, the nature of demand and supply is so dynamic that real-time updating of the central data-warehouse is required, and this is a Stage 4 characteristic. The most typical type of business environment where this dynamic real-time need is in place is an environment where promotional activity is highly prevalent. In these settings, demand-supply balancing happens in real time, and this requires a data warehouse to be updated in real-time to facilitate this balancing.

System Infrastructure

The final theme in the systems dimension is what we refer to as *system infrastructure*, and the difference between Stage 1 and Stage 4 is something of a matter of judgment—is the company doing enough to provide the hardware, software, and IT support to allow the demand forecasting function to operate effectively? For example, at one company we audited, the forecasting process required its sales teams, located across the world, to interact directly with the forecasting system on a monthly basis to adjust their baseline statistical forecasts. However, the sales teams found that they were waiting literally hours for their screens to load on their computers, and as a result, were not submitting forecasts. This company had "inadequate" infrastructure, and was at Stage 1. At a different company, the forecasting software was three upgrade levels behind at the time of our audit, and many capabilities were not being utilized because the upgrade was far down on the priority queue for IT support. This company was also at Stage 1. At still another company, the demand planning team had a dedicated IT support person, whose full-time job was to keep the system up-to-date with upgrades, and turnaround for requested enhancements was literally days, rather than weeks or months. This company had considerable hardware capacity for a very compute-intensive forecasting environment (hundreds of thousands of monthly forecasts at the SKU by location level of detail). This company had "superior" infrastructure, and was at Stage 4. Other companies have been somewhere in-between, with some having "minimal" infrastructure (Stage 2) and others having "acceptable" infrastructure (Stage 3).

Summary: How Companies Can Improve on the Systems Dimension

To review, the systems dimension answers the question, "What is the state of information technology support for the demand planning process?" Functional integration is about culture; approach is about process; systems is about tools. If you have highlighted too many Stage 1 or Stage 2 characteristics in the systems dimension, here are some steps that you can take to move your demand forecasting systems in the direction of being world class:

- **Invest in integration.** This is not to say that the only demand forecasting systems that companies should buy are those that are part of integrated supply chain suites. Some fine demand forecasting systems are "best-in-breed," and they can integrate well into the firm's overall IT architecture. However, when making forecasting system decisions, companies should pay considerable attention to the extent to which the forecasting system integrates into the company's IT backbone, how it accesses the company's data warehouse, and how it exchanges information with other corporate systems.

- **Provide users with performance metrics.** Because forecasting is a management process, performance metrics are critical, and without access to these performance metrics by people who need them, the process cannot be managed adequately.

- **Make sure the forecasts are credible by ensuring data integrity.** Few things can undermine the integrity of a forecast any faster than to have a user say, "Your data is not correct!" Integrating the forecasting process into the firm's data warehouse strategy is critical.

- **Invest in infrastructure.** Companies find that investing in forecasting processes, and then skimping on services such as IT support and version upgrades, is "penny-wise and pound-foolish."

One final word of caution concerning systems: I've stated quite strongly that "systems are not silver bullets." In my experience, it is very common for a company to be unsatisfied with its forecasting efforts, and immediately try to solve the problems by buying and installing a forecasting system. Although not a scientifically proven hypothesis, my guess as to why this happens is that it is the easiest path to follow. When I say "easy," I don't mean that no work is involved in installing a system. Far from it. But in many companies, it's a kind of work that people are familiar with. It lends itself to project management, to Pert charts and Gantt charts, and assignment of responsibility and deadlines and status reports. Changing culture is much more nebulous, and doesn't lend itself nearly as well to familiar project management frameworks. Companies are often good at project management, but not good at change management. Installing a new culture is more challenging than installing a new system. Unfortunately, the culture change is usually much more impactful than the system change.

Performance Measurement

The final dimension to be examined in our discussion of world-class demand forecasting is *performance measurement*. This dimension answers the question, "How appropriately is demand forecasting performance measured and rewarded?" Functional integration is all about culture; approach is all about process, and systems is all about tools. Performance measurement ties these dimensions together. This final dimension is all about utilizing *tools* to measure *processes*, which influences *culture*. The performance measurement dimension has only two themes, as shown in Table 7-4: how performance is measured, and how performance is rewarded.

How Is Performance Measured?

At Stage 1 companies, forecasting performance is not measured. When we work with a company, it is quite evident when the company is at Stage 1 on this first theme. Questions about forecasting accuracy are met with blank stares, or with answers like "Well, we certainly could do better," or "Well, I think our accuracy is somewhere around 75%" (unless the person being interviewed is a "victim" of bad forecasts, in which case the answer is more like "Our accuracy is somewhere around 25%"!) Any reader who has made it this far in this book should recognize that not measuring forecasting performance is a problem!

Table 7-4 Performance Measurement

Theme	Stage 1	Stage 2	Stage 3	Stage 4
How is performance measured?	Performance is not measured.	Accuracy is measured, primarily with MAPE.	Accuracy is measured with MAPE, and bias is graphically depicted with PE graphs.	Multi-dimensional metrics of performance are used. Accuracy and bias are tied to supply chain metrics.
How is performance rewarded?	Performance is not tied to any measure of accuracy.	Performance is tied to accuracy for forecasters only.	Performance is tied to accuracy for everyone involved in the demand planning process.	Supply chain metrics (for example, inventory) and customer service metrics (for example, fill rates) are shared across the enterprise.

At Stage 2 companies, accuracy is the focus of all performance measurement efforts, and the most common metric is Mean Absolute Percent Error (MAPE), or Mean Absolute Percent Accuracy, which is 1-MAPE. As discussed in length in Chapter 6, MAPE is the most common accuracy metric used, and it is completely appropriate, if measured correctly, as a "scorecard metric." It is useful for tracking individual or group forecasting performance over time, and it is an appropriate surrogate for demand volatility, and thus useful for calculating safety stock levels. However, the main problem with a company being at Stage 2 is that bias is not measured. MAPE is an excellence scorecard metric, but not a very good diagnostic metric, and without the effort being put into examining bias, diagnosing systematic forecasting problems is often difficult.

Stage 3 companies add examination of bias into their routine performance measurement processes. Bias, which is most commonly measured using Percent Error (PE), and which is most effectively depicted graphically, is an excellent way to detect systematic forecasting problems. As discussed in Chapter 6, PE graphs can quickly point out individuals who have "other agendas" in regard to their forecasting responsibilities, such as salespeople who underforecast to influence sales targets, or brand managers who overforecast to increase their advertising budgets. Stage 3 companies, then, utilize both the scorecard capabilities of MAPE, and the diagnostic capabilities of PE graphs.

Finally, Stage 4 companies move beyond the "process metrics" such as MAPE and PE, and also consider the "outcome metrics" such as inventory turns, customer fill rates, and expediting costs, to help guide strategic decision making. As discussed in Chapter 6, *no one buys or sells stock in a company because they're good at forecasting*. Rather, overall corporate performance is judged by outcome metrics. World-class companies use process metrics such as forecast accuracy to make good decisions about which products need to be carried in inventory in which locations, so as to balance cost and customer service needs.

How Is Performance Rewarded?

Stage 1 companies fail to follow the management maxim of "What gets measured gets rewarded, and what gets rewarded gets done." At these companies, no effort is made to reward individuals for creating, or contributing to, accurate, unbiased forecasts. At one company we worked with, management was at least beginning to take the journey toward accountability for those who contribute to the forecasting process. At this company, salespeople were paid a bonus if they submitted their forecasts *on time*. Although this is arguably a "crawl before you walk" strategy, our team still judged this company to be Stage 1 on this theme.

Stage 2 in this theme is the most common place for a company in our database to be. Here, companies do in fact reward excellence in forecasting, but limit those rewarded to the individuals who work in the forecasting function. As discussed in Chapter 6, these rewards can come from either achieving a pre-determined accuracy target for their assigned product or customer forecasts, or they can come from period-over-period improvements. Although Stage 2 here is clearly superior to Stage 1, problems still remain. Forecasters often rely heavily on the insights contributed by sales, marketing, brand management, and senior executives, and if those individuals are not incentivized to provide accurate, unbiased forecasts, then the forecasters don't have the tools necessary to achieve the best results possible. Stage 3 companies overcome this problem by providing incentives for everyone in the company who participates in the forecasting process. Individuals at Stage 3 companies, who work in sales, marketing, brand management, and even senior management, find that some element of their performance plan contains achievement of forecast accuracy goals.

Movement to Stage 4 in this theme requires that companies move beyond the process metrics, and work to influence the outcome metrics. Stage 4 companies have cross-functional metrics, and rewards, shared across the enterprise. For example, at one company in our database, individuals in the sales organization became accountable not only for forecast accuracy, but for finished goods inventory levels. This level of accountability helped the sales teams see that forecast accuracy in and of itself was not nearly as important as was the effect of that accuracy on important corporate goals. Implementation of this reward strategy reduced the amount of upward bias in the forecasts coming out of sales. In this case, "what got rewarded got done!"

Summary: How Companies Can Improve on the Performance Measurement Dimension

The final dimension of demand forecasting excellence is very straightforward but very important. I offer two simple statements to guide process improvement in this area:

- Measure performance
- Reward performance

Seems simple, right? Of course, if it were simple, companies would do it, and many don't. In our experience of working with dozens of companies, measuring and rewarding the forecasting personnel is fairly easy. It is, however, quite rare for companies to reach Stage 3 in this dimension, and begin to include other functions such as sales, marketing, brand management, and senior leaders into the "measure and reward" process. Performance measurement describes how well companies utilize tools to measure processes that ultimately influence culture. Recall that culture represents 60% of the challenge to forecasting and DSI excellence, and measuring and rewarding performance is perhaps the most compelling tool available to change that culture.

Summary of World Class Forecasting

This chapter is far too long to neatly summarize in one short section. Rather than simply re-stating the main points from each of the dimensions of forecasting excellence that have been described, I summarize this discussion of world-class forecasting by summarizing the points made in an article published several years ago by our research team called, "The Seven Keys to Better Forecasting."[10] Although

---

[10] Moon, Mark A., John T. Mentzer, Carlo D. Smith and Michael S. Garver (1998), "Seven Keys to Better Forecasting," *Business Horizons*, (September–October), 44–52.

much has been added to our understanding of what constitutes world-class forecasting since this article was written, it does provide many useful insights about forecasting excellence, and as such, it constitutes a good way to summarize the discussion in this chapter. "The Seven Keys to Better Forecasting" (not to be confused with the Seven Deadly Sins!), are

1. **Understand what forecasting is, and what it is not.** Forecasting is a management process; it is not a software program. Forecasting is a best guess about what future demand is likely to be; it is not a plan or a goal.

2. **Forecast demand and plan supply.** The focus of the demand forecasting process should be to try and anticipate future *demand*, or what customers would buy from us if they could. Too many companies limit their forecasting efforts to an estimate of what future *sales* will be. A strategic view that drives the supply chain, as well as the business as a whole, should be one that attempts to forecast actual demand.

3. **Communicate, coordinate, collaborate.** Important insights about future demand are available from a variety of sources: history, the judgment of experienced people, even customers. Creating a culture, and a process, where insights and information are freely shared, is critical to world-class forecasting.

4. **Eliminate "islands of analysis."** Excel continues to the be the most popular software tool for demand forecasting, and with no disrespect toward the Microsoft Corporation, it is the wrong tool for this job. The "spreadsheet mania" seen at many companies inhibits effective cross-functional, multi-divisional, and global demand/supply integration. Although "tools" only drive 10% of DSI excellence, it's a critical 10%, and getting the tools aligned is critical.

5. **Use techniques wisely.** Demand forecasters have a variety of forecasting techniques at their fingertips, and Chapters 3 and 4 are dedicated to these techniques. There are both quantitative

technique (time-series, regression, causal modeling) and qualitative techniques (jury of executive opinion, Delphi methods, salesforce composite, customer forecasts). Using these techniques "wisely" means applying them in the circumstances and sequences that makes the most sense.

6. **Make it important.** For forecasting and demand planning to be effective, individuals must take it seriously. Resources must be made available, both human and financial. A company doesn't see forecasting as important when it has 40,000 SKUs to forecast each month, and only three demand planners, or when a company won't invest in the system infrastructure needed to do the job well, or when a company will not create reward structures for everyone who participates in the process. To make it important, the company has to. . .

7. **Measure, measure, measure.** "What gets measured gets rewarded, and what gets rewarded gets done." Many of the problems in forecasting processes are fixed only when forecasting performance is measured. Forecasting is a management process, and like any management process, it cannot be managed well if it is not measured.

This concludes our discussion of what constitutes world-class forecasting. To bring this book full circle, we return to the topic first introduced in the book's title and explained in great detail in Chapter 1, "Demand/Supply Integration." Chapter 8 returns to the DSI topic by discussing how to best incorporate the excellent demand forecasts that have been created by following the ideas in Chapters 2–7, by effectively managing the demand review.

8

Bringing It Back to Demand/Supply Integration: Managing the Demand Review

Chapter 1 focused on the "super-process" of demand/supply integration, or DSI. Chapters 2–7 examined the subprocess of demand forecasting, which along with a variety of other subprocesses such as supply planning, inventory planning, and financial planning, make up the super-process of DSI. This concluding chapter focuses once again on demand/supply integration, but, the discussion centers on the demand review. One way to think about the demand forecasting process is that it is a month-long exercise in preparation for the demand review. All the steps discussed in the book so far, from statistical forecasting, to qualitative forecasting, to performance measurement—all are either foundational capabilities that must be in place, or specific pieces of the subprocess that lead up to the demand review. This chapter presents the typical process flow that leads up to that demand review, as well as the most effective way to conduct the actual demand review meeting, paying considerable attention to that step I call "gap analysis." This critical step transforms DSI from a tactical exercise in supply chain planning to a strategic element of the overall business planning process in the firm.

Figure 8-1 shows a graphical representation of the Demand Forecasting Process flow, which consists of three distinct phases. Phase I is perhaps the most laborious and time consuming. The outcome of this phase is the initial forecast, which follows from the consolidation of various subprocesses that have been discussed in previous chapters. In Phase II, the demand forecaster identifies the gaps between the initial forecast and the overall goals of the firm, and creates a series of gap-closing strategies in preparation for the demand review. Phase III is the actual demand review meeting. The following sections describe the work that needs to be done to effectively complete each of these phases in the demand forecasting process flow.

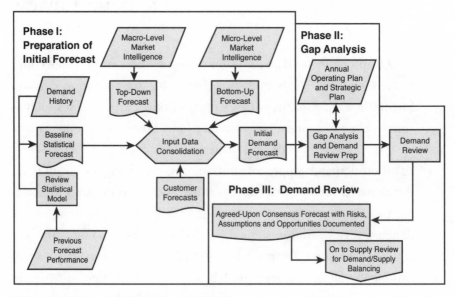

Figure 8-1 Demand Forecasting Process flow

Phase I: Preparation of Initial Forecast

The entire process usually begins with the step labeled "Baseline Statistical Forecast" in Figure 8-1. I say that it usually begins with this step because as Chapter 3 detailed, some relatively rare situations exist where an analysis of historical demand patterns is not particularly useful. Aside from these situations, the baseline statistical forecast requires access to demand history. I have spoken in various places throughout this book of the importance of using *demand history* as the source of the *demand forecast*. As discussed in Chapter 7, world-class companies construct their demand history using three separate pieces of information: shipment history, adjustments for backorders, and adjustments for unrecognized demand, or lost sales. Frequently, creation of this demand history data file requires a monumental effort that involves not only system enhancements, but behavioral changes by people who work with customers. Both sales and customer service must be trained, and incentivized, to document those instances where customers were ready but unable to buy, because the product or service was not available at the time or place required by the customer. Individuals in these departments require access to this demand history file so that these lost orders can be documented. All this data—shipment history, backorder adjustments, and lost order records—should be stored and professionally maintained in the firm's data warehouse. Refer to Figure 2-2 or Figure 7-1 for a refresher on the appropriate system infrastructure that supports demand forecasting.

After accessing the demand history, analysts then can apply the procedures described in Chapter 3 to "look in the rear-view mirror" for patterns that might exist in that historical demand, and then project those patterns into the future. However, before applying whatever statistical models might have been used in the past, reviewing those statistical models to ensure that they are still of value is important. Chapter 6 discussed how performance measurement techniques can be used as diagnostic tools to evaluate the usefulness of various statistical models. Examples were offered of percent error charts that revealed flaws in the forecasting techniques that had been applied; also before finalizing the statistical forecast, the analyst should look at previous periods' performance metrics to identify models that should be adjusted or rethought.

After the baseline statistical forecast is created, the analyst must begin to consolidate the various sources of data that are used to answer the question, "Will the future look like the past?" Various inputs are used to answer this question. These inputs, which the forecaster must consolidate, include a top-down forecast, which is created with macro-level market intelligence as described in Chapter 5. As noted in that chapter, and in Chapter 7 during the discussion of the "Approach" dimension of forecasting management, the most effective process is one that encompasses both a top-down and a bottom-up perspective. Recall that a top-down perspective is one where an estimate of industry demand is combined with an estimate of market share to arrive at a forecast of demand. Macro-level information is needed to create that top-down perspective. Recall from Table 5-1 (Micro versus Macro Market Intelligence) how market intelligence can help to inform that critical top-down demand forecast. As discussed in Chapter 5, forecasters often struggle to include this macro-level market intelligence in their demand forecasting process. However, including this step in a process flow such as that depicted in Figure 8-1 helps to remind forecasters of the importance of looking at macro-level information on a regular basis, and using that information to continuously analyze and document those critical assumptions that underlie the forecast.

Another input that must be included in the "Input Data Consolidation" step is a bottom-up forecast. Chapter 4 covered the importance of qualitative judgment, which is usually gathered from sales, marketing, and product management in a manufacturing context, and merchandising in a retail context. This qualitative judgment constitutes the key element of the micro-level market intelligence discussed in Chapter 5. Insights about customers that come from sales, as well as information about promotional activity that comes from marketing, product management, or merchandising, is critical in creating this bottom-up forecast. The final piece of input that analysts must consolidate during this phase is customer-generated forecasts, which Chapter 5 also discussed. That chapter discussed the best way to choose which customers should be providing forecasts, and some of the risks and opportunities that are involved in using customer-generated forecasts. In many situations, though, these direct customer insights are extremely useful. Thus, the "Input Data Consolidation" step in Figure 8-1 consists of data from the baseline statistical forecast, the top-down forecast generated through macro-level market intelligence, the bottom-up forecast generated through micro-level market intelligence, and customer-generated forecasts.

At this point in the process is where competent demand forecasters look completely different from excellent demand forecasters. Competent demand forecasters are capable of pulling together this information and compiling it into a database or spreadsheet. Excellent demand forecasters are able to take the forecasts created from the different perspectives—statistical, top-down, bottom-up, customer-generated—and interpret the biases, understand the various agendas, evaluate the different levels of quality, apply their own intuition and insight, and create an initial demand forecast that will be ready for the next step in the process—gap analysis.

Phase II: Gap Analysis

The entire purpose behind all the work that is completed in Phase I of the demand forecasting process depicted in Figure 8-1 is to create the best, most accurate, most credible forecast of demand in future time periods. It is intended to be a dispassionate assessment of the level of demand in the marketplace for the firm's goods and services. Arriving at this place takes a lot of effort by a lot of people—but it's not the end of the job, because, as discussed in Chapter 1, the case might be that the best, most accurate, and most credible forecast of demand results in the conclusion that the firm will not achieve its objectives. If that is the result of the demand forecasting process, then identifying the gaps becomes the responsibility of the demand forecasters, as well as preparing gap-closing strategies that can be discussed at the demand review. This section discusses the concept of gaps and tries to bring clarity to the cause of those gaps, as well as identifies some of the possible gap-closing options.

Chapter 1 discussed the difference between *forecasts* and *goals*. Recall that a forecast is the best guess about what will actually happen, given a set of assumptions. A goal is the outcome that the firm hopes will happen. Goals can be expressed in different ways. A firm can have market share goals, margin goals, inventory goals, cash-flow goals, revenue goals, or any other of a variety of goals. Many organizations have an overarching set of goals, typically financial in nature, which is stated in annual or quarterly "buckets." Usually referred to as the Annual Operating Plan, or AOP (even though it is really a goal and not a plan), this "master-goal" often forms the foundation for all goals established by the firm. It is commonly the case that the *forecast*—what we think will actually happen—falls short of the *goal*—what we hoped would happen, and what we planned our business to be able to have happen.

So what are the consequences of failing to generate enough demand to achieve the goals expressed in the AOP? The two primary consequences are financial and operational. The financial consequence is that for publicly traded firms, investors tend to value the firm based upon their expectations of the firm's performance. Regardless of the firm's strategy for communicating expectations to investors, typically the case is that when actual performance fails to reach stated goals, investors won't be happy, and the stock price might suffer. The operational consequences are that from a planning perspective, the firm will typically acquire enough supply capacity to allow it to deliver the goods or services that must be sold in order to achieve the AOP. Thus, if insufficient demand exists in the marketplace to actually generate the revenue stated in the AOP, then unused capacity might result. Raw material and work-in-process (WIP) inventory might stack up, workers might need to be laid off or furloughed, and investment in fixed cost capacity expansion might be wasted. In other words, across a variety of dimensions, failure to generate enough demand to achieve the AOP goals is not a good thing.

Unfortunately, what often happens in this situation is what I described in Chapter 1 as the most insidious aberration to an effective DSI process—*plan-driven forecasting*. When the forecast fails to be as high as the AOP, the forecast is simply raised up to the point where it is consistent with the AOP, and the firm deludes itself into thinking that everything is okay. This is insidious because it removes all credibility from the forecasting process. "Customers" of the forecast—those procurement planners, production planners, inventory planners, transportation planners, financial planners, and so on—begin to ignore the forecast because they don't believe it is based on actual demand in the marketplace. This is why the gap analysis phase of the demand forecasting process flow is so important. Without the disciplined analytical activities that accompany this gap analysis, the firm not only runs the risk of failing to achieve its objectives, accompanied by the resulting consequences, but it also runs the risk of removing credibility from the forecasting process as a whole.

Three separate steps are involved in an effective gap analysis. The first step is to examine the assumptions underlying the AOP. Although the possibility exists that the AOP was determined by something as simplistic as "Our plan for the upcoming year is to increase everything by 10%," one would hope that a more comprehensive analysis was done, with assumptions underlying those analyses. Typically, assumptions that underlie the AOP include the following:

- **General business climate.** Macroeconomic assumptions can include statements about economic growth, unemployment rates, interest rates, or whatever general business indicators are relevant for the business being planned.

- **Market share.** When the firm is making overall business plans, it needs to make assumptions about its market share in different markets. General business climate assumptions will inform overall industry sales predictions, but market share assumptions are needed when the firm is planning its expected long-term demand.

- **Industry growth.** Beyond the general business climate, the industry in which the firm competes might grow or contract at a different rate, or in a different direction, than the general economy. Assumptions must be made about overall industry growth when planning the business.

- **Competitive activity.** Assumptions about competitive activity will underlie the market share assumptions noted previously. Market share is likely to remain stable if neither the firm nor its competitors do anything different than they've done before. However, in most cases, neither the firm nor its competitors will remain static.

Each firm, and each industry, has its own set of assumptions that underlie their AOPs. The more completely these assumptions are documented, the easier is the job of demand forecasters when doing their gap analysis. If the forecast does not reflect the level of demand found in the AOP, there are really only two possible reasons: either the firm's performance has not reached expectations, or the industry-level assumptions underlying the AOP have not in fact occurred as they were planned. Determining which of these root causes is the real reason behind the gap is critical. If the industry-level assumptions have not occurred as planned, there might be little that can be done, at least in the short run. However, if the gaps are caused by firm performance issues, then gap-closing actions are probably available for consideration.

The second step involved in gap analysis is to document the magnitude, and the level, of the gaps. Several different categories of gaps can exist between the forecast and the AOP, and an understanding of them can help to guide the demand forecasters in their recommendations for gap-closing strategies. These different categories are

- **Timing gap.** In some cases, demand is likely to materialize, but the timing of the demand is not consistent with the expectations that underlie the AOP. For example, AOP assumptions might include incremental demand associated with new product launches. Having new product launches be delayed, for any number of reasons, is not uncommon. In this case, the demand assumptions might still be valid, but because of the launch delay, the forecast will not match up with the AOP. Another timing gap might revolve around project-based businesses. Again, the assumption behind the magnitude of demand might be valid, but customers who have awarded large projects to the firm might be experiencing delays in the implementation of these projects, and this might affect the timing of their actual purchases. This might be reflected in the demand forecast. In either of these cases, there might be no need for the demand forecaster to suggest any gap-closing strategies, but rather, to simply update others in the firm about these timing issues.

• **Volume gap.** In some cases, the overall volume of demand might be reasonably close to the AOP goal, but the mix of SKUs, or even brands, that constitute the overall volume, might be highly uncertain. This uncertainty can have a substantial impact not only on revenues, but also on profits. As discussed in Chapter 2, when a company forecasts at the SKU level, there will inevitably be more error, because lower levels of the forecasting hierarchy usually have considerably more variability of demand.

• **Regional gap.** It is frequently the case that some regions of a company's market area will experience demand in line with expectations, while others will not. For example, a company might create a forecast for demand in Germany that is consistent with the AOP, while during the same period, demand in Spain or Greece would be far below expectations due to continuing economic woes in those countries. In those cases, suggesting gap-closing strategies would be very helpful for the demand forecaster, such as to increase demand in regions where economic conditions are more favorable.

- **Customer gap.** Just as situations might exist where one region is meeting planned targets while another is falling short, there might also be certain customers whose demand levels are meeting expectations while other customers are buying at far less than anticipated levels. Gap-closing strategies might be available to increase demand at some customers, because demand is lagging at others.

The bottom line then, from this discussion, is that understanding the source of the gap between AOP and forecasted demand is critical. Without such understanding, any gap-closing strategies are likely to be misdirected and ineffective.

The final step in the gap analysis process is in preparation for the demand review, to prepare a set of alternative gap-closing strategies to present for consideration in that meeting. One important element is that this stage of the process is *demand focused*. In other words, the gaps that have been identified up to this point are gaps between *what customers would buy from us if they could*—remember, that's our definition of *demand*—and what our firm had planned for in its Annual Operating Plan. No discussion should occur during the demand review of gap closing strategies related to *supply*. The focus at this point is on the question of *how can the firm influence customer demand to bring demand shortfalls into alignment with the firm's overall goals?* To answer this question, I return to a discussion from Chapter 1. When demand is falling short of expectations, a variety of "levers" can be pulled. Some of these levers are very short term–oriented, such as

- **Promotional activity.** In many companies, additional demand can quickly be acquired through sales promotion efforts. In consumer packaged goods (CPG) companies, either trade or consumer-based promotions can have a dramatic, although short-term, effect on demand. In business-to-business firms, promotional activity might take the form of salesperson incentives to increase demand in certain product categories to certain customers or channels. Demand-side executives should always keep in mind the fact that the demand spikes that often accompany these promotions can be highly disruptive to the supply chain, creating peaks and troughs of demand that can be quite costly.

- **Pricing actions.** Because most demand curves are downward sloping (at least I think I remember that from my economics courses long ago), firms can usually expect that a price reduction increases demand, and a price increase decreases demand. The amount of the expected demand change is, of course, determined by the buyer's price elasticity of demand. The firm also must take into consideration any strategic implications of pricing actions, especially on brand reputation. For example, I wouldn't expect that executives at a company such as Rolex would approve of a price reduction as a way to close any gaps between forecasted demand and the AOP. Such an action could negatively affect the brand's reputation, and potentially lower the consumer's reference price for that product.

- **New Product Introduction (NPI) timing.** In some cases, NPI timing can either be delayed, or accelerated, to create a gap-closing strategy. For example, if an "old" product that is being replaced by a new product is seeing volume declines that are more rapid than originally expected, and if the new product is ready to introduce earlier than planned, closing a gap is possible by changing the timing of the new product introduction.

Of course, these short term–oriented gap-closing strategies will sometimes have a "rob Peter to pay Paul" effect. If a new product is introduced earlier than anticipated, then the demand might be shifted to an earlier period, but overall demand levels might not change. If immediate demand is increased through a price promotion, it might mean that either business customers or consumers will "load up" and not buy in future time periods. More long-term strategies for increasing demand up to the levels targeted in the AOP might include expanding to new markets, introducing new brands that might appeal to underserved markets, or using new overall marketing mix strategies designed to revive mature or declining markets.

Thus, the task of the demand forecaster, in preparation for the demand review, is to prepare the initial demand forecast (Phase I of Figure 8-1), identify gaps between forecasted demand levels and the targets articulated in the Annual Operating Plan, and identify possible gap-closing strategies that can be reviewed during the demand review itself (Phase II).

Phase III: Demand Review Meeting

As stated at the beginning of this chapter, you can think of the entire demand forecasting process as preparation for the demand review meeting. When viewed in the context of the entire DSI super-process, the demand review is typically the first major step. It is here that the demand side of the enterprise (sales and marketing in a manufacturing context and merchandising in a retailing context) passes along to the supply side of the enterprise its best guess about the level of demand that their efforts will generate over the upcoming planning period. At its best, the demand review is seen as a hard commitment on the part of the commercial team to deliver that stated level of demand to the company. This demand forecast then drives the supply team to finalize its plans for all the supply chain components (transportation, production, procurement, and so on) that are needed to support the level of demand to which the commercial team has committed. It also drives the financial team to acquire the capital needed to support this level of demand, and to report to the company's owners, whether those are shareholders or outright owners, on the financial outcome they can expect. This is why it warrants all the work described throughout the bulk of this book!

The agenda for the demand review meeting should include a number of items. One should be a review of previous month's performance and assumptions. This agenda item should serve only as the first step, and not as the focus of the meeting. One of the typical problems that I've observed in DSI processes is that the meetings are too focused on "How did we do last month?" rather than on "What decisions should we make *now* in anticipation of *future* demand?" Still, a review of past performance, along with a review of the status of documented assumptions from previous months, is an appropriate starting point for the demand review meeting.

Another agenda item is to review the initial forecast, *by exception*. Recall our discussion earlier in this chapter, where the point was made that this initial forecast will consist of an amalgamation of several inputs: the statistical forecast; the bottom-up forecast, which is created using micro-level market intelligence; the top-down forecast, which is created using macro-level market intelligence; and possibly customer-generated forecasts. In most companies, thousands or even tens of thousands of initial forecasts are created each month, depending on the appropriate forecasting level employed. Obviously, a demand planner cannot review each initial forecast, nor can these reports be discussed at the demand review. Two strategies are employed to manage this complexity. The first strategy is to establish exception rules, which drive the decision of whether, and how, to discuss the forecast at the demand review. Table 8-1 shows an example of such a set of exception rules. This table assumes that an "ABC" classification scheme is in place, where the most important products are classified as "A" products, mid-level products are classified as "B" level products, and low-importance products are classified as "C" level products.

Table 8-1 Example of Exception Rule for Demand Review

Product Classification	Discussed at Demand Review If:
A (top 10% of products by volume or revenue)	All A level products are covered at the demand review.
B (middle 70% of products by volume or revenue)	B level products are only discussed if forecast error (MAPE) for the past 3 months falls above a previously determined threshold; for example, 30%.
C (bottom 20% of products by volume or revenue)	C level products are only discussed if forecast error (MAPE) for the past 3 months falls above a previously determined threshold; for example, 50%.

The MAPE figures that are included in Table 8-1 are used as illustrations—the actual threshold that a company should use is highly idiosyncratic to each company. The point is, though, to establish some sort of *decision rule* to drive the decision about which forecasts to discuss.

The second strategy used to manage this product complexity is to make forecasting decisions, not at the SKU level, but at a higher level of aggregation in the forecasting hierarchy. Recall from Chapter 3 the discussion of how SKU-level forecasting can often be problematic because of the excessive variability that is experienced at the SKU level. I made the point there that many companies forecast at the product family level, because discernible patterns often exist at that higher level of aggregation that don't exist at lower levels. When companies take this approach, they often manage the complexity of thousands of SKUs by creating exception rules at the product family level, not at the SKU level. Then, all A level product families are discussed at the demand review, and B and C level product families are discussed only if they fail to meet a pre-determined accuracy threshold at the product family level.

You can find an example of some of these strategies in an article by Mentzer and Schroeter.[1] The company they worked with was Brake Parts, Inc., a manufacturer of aftermarket automobile brake systems and parts. The daunting task faced by this company's forecasting team was a monthly workload of more than 600,000 SKU-by-location forecasts. Obviously, something needed to be done to manage this, because no team of forecasters would be able to analyze 600,000 forecasts, and no demand review would consist of discussing and reaching consensus on 600,000 forecasts! Their solution was to utilize technology and rely on their statistical forecasting system to grind through all those forecasts, and use exception rules to identify for the demand forecasters those specific products that required human intervention and thought. They also managed their demand review by product family, and again used exception rules to drive their decision about which products to discuss at the demand review meeting. Their goal was to create a system that would effectively forecast demand for these products, and only require human beings to examine or discuss a maximum of 1,000 of the 600,000 products they forecasted each month.

The next agenda item in the demand review is to discuss significant results from the portfolio and product review, focusing on high-impact new product introductions scheduled for the near-term, and decisions made concerning any significant SKU reductions and their effect on demand for other products. Following these discussions, the demand forecasting team should be prepared to present the results of their gap analysis. Articulation of the anticipated gaps, an analysis of the type of gap involved, a presentation of possible demand-side gap-closing strategies, and discussion among decision-capable participants should all occur. To return to a critical point that first introduced in Chapter 1: key, decision-capable representatives from the demand side of the enterprise should attend the demand review

[1] Mentzer, John T. and Jon Schroeter (1993), "Multiple Forecasting System at Brake Parts, Inc.," *Journal of Business Forecasting*, (Fall), 5–9.

meeting, including product or brand marketing, sales, customer service, and key account management. This discussion of gap-closing strategies is the reason that "decision-capable" individuals must be present at the demand review. Even before demand-supply balancing takes place at the supply review stage, *decisions need to be made at the demand review about which demand "levers" should be pulled to bring demand and supply into alignment.* As noted in the best practices discussion in Chapter 7, Stage 2 companies might have a formalized DSI process in place, but they often fail to have decision makers in attendance at the key meetings. Without these decision-capable individuals in the room, the demand review often reverts to a discussion about "why we didn't make our numbers last month." Stage 3 and Stage 4 companies—those who are best in class—have the key players in attendance at all the critical meetings, where decisions can be made and communicated to all other relevant parties.

After discussing these gap closing strategies, and making the decisions, having those in attendance make a statement of consensus is critically important. As discussed in various places in this book, a spirit of consensus defines the optimal culture for DSI, and such a statement of consensus at the demand review meeting ensures that all participants have "bought in" to the decisions that have been reached. I've attended formal demand review meetings where the accepted "protocol" is that at the end of the meeting, the meeting chair literally points to each person in the room, and asks for a verbal statement of support that the numbers that have been discussed are the numbers the group will commit to, and that the gap-closing decisions that have been reached have the support of the group. Sometimes, meeting attendees will not be comfortable voicing that support, and further discussion results. But by the end of the meeting, all important players have gone on record in support of the group's decisions.

The output, then, of the demand review meeting is the consensus forecast of demand, and the agreed-upon gap-closing strategies. But that's not enough. Another, equally important output of the demand review is a clear statement of the assumptions that underlie the forecast, and any risks and opportunities associated with the forecast. Chapter 5 covered these assumptions in some detail, and the collection and interpretation of market intelligence forms the basis of these internal, and external, assumptions.

Conclusions

This brings us to the end. The consensus demand forecast that comes out of that demand review now goes off to inform the rest of the demand/supply integration (DSI) process. It goes to the supply review meeting, where the supply side of the business will match it up against their capacity forecast, balance total forecasted demand with total forecasted supply, and identify issues that need to be resolved at higher levels in the firm. It then goes to the reconciliation meeting, where the financial community of the firm gets actively involved, dollarizes all the decisions made at earlier meetings, and resolves any issues that can be resolved. Finally, it goes to the executive DSI meeting, where the firm's leadership team makes sure that the plans that have been agreed upon to capture the identified demand are in alignment with the goals and strategic direction of the enterprise.

And then, you do it all over again.

As I conclude, allow me to make some summary comments, all of which have been made elsewhere in this book, but which deserve one more mention at its conclusion. These represent random neuron firings, and are in no particular order of priority.

- Because a forecast is a guess about the future, it will *always* be wrong. The challenge is to make it the least wrong that it can be.
- No one buys stock in a company because that company is good at forecasting. Forecasting is only important, or interesting, or

worth the effort, if it leads to good business decisions that serve customers, enhance revenue, and reduce costs.

- Statistical forecasting is a necessary, but insufficient, step in a good demand forecasting process. Remember, if you only look in the rear-view mirror, you are likely to get hit by a truck.

- Sales and marketing or merchandising must participate. Period.

- Senior executives must buy into DSI as a way to run the company, and put their money where their mouths are. Without executive support, both financial and emotional, DSI will fail. Period.

- An organization's culture is far more important to the success of DSI, and good demand forecasting, than any process flowchart or any piece of technology.

And with that, I now conclude. I hope that all of your forecasts are accurate, and that all of your businesses are successful.

Index

A

aberrations in DSI, 14-18
  DSI as tactical process, 16
  lack of alignment with sales/
    marketing, 17
  plan-driven forecasting, 14
accountability, 27
  functional integration, 181-183
  measuring, 144
accuracy. *See* MAPE (Mean Absolute
  Percent Error)
AOP (Annual Operating Plan), 224
  gaps between forecast and AOP,
    226-228
  underlying assumptions, 225-226
apparel industry, 1-2
approach
  forecasting hierarchy, 194-196
  forecasting point of view, 189-192
  forecasting true demand, 192-194
  improving, 199-201
  incorporation of qualitative input,
    197-199
  statistical analysis, 196-197
audits, 173-174
available data, nature of, 45-47
averages
  moving average, 68-72
  simple average, 63-68

B

Benchmark Studies, 171-173
bias
  identifying, 149-154
  in jury of executive opinion, 105-106
black-box forecasting, 79-80
Boeing Corporation
  customer base, 44-45
  qualitative forecasting, 95-96
bottom-up forecasting, 121-122,
  190-191
Brake Parts, Inc.,233
business climate, 225
business planning
  compared to demand forecasting,
    33, 183-184
  demand plans, 10
  financial plans, 11
  operational plans, 10

C

calculating
  MAPE (Mean Absolute Percent
    Error), 155-157
    unweighted MAPE, 160
    weighted aggregate MAPE,
    160-163
    weighted MAPE, 159-160
  moving average, 68-72
  percent error, 145-149
  return on shareholder value,
    165-168
  simple average, 64-68
causality versus correlation, 90
change in culture, 28-29